Steroid Converting
Enzymes and Diseases

Steroid Converting Enzymes and Diseases

Editors
K. Fotherby · S. B. Pal

W DE G

Walter de Gruyter · Berlin · New York 1984

Editors

K. Fotherby, Ph. D., F.R.I.C.
Department of Steroid Biochemistry
Royal Postgraduate Medical School
University of London
Ducane Road
London W12 OHS
United Kingdom

S. B. Pal, D. Phil., Dr. rer. biol. hum., M. I. Biol.
Universität Ulm
Department für Innere Medizin
Steinhövelstraße 9
D-7900 Ulm
F. R. of Germany

CIP-Kurztitelaufnahme der Deutschen Bibliothek

Steroid converting : enzymes and diseases /
ed. K. Fotherby ; S. B. Pal. - Berlin ; New York : de Gruyter, 1984
ISBN 3-11-009556-4

NE: Fotherby, Kenneth [Hrsg.]

Library of Congress Cataloging in Publication Data

Main entry under title:
Steroid converting enzymes and diseases.
Bibliography: p.
Includes index.
1. Steroid hormones--Metabolism. 2. Steroid
hormones--Metabolism--Disorders. 3. Enzymes.
I. Fotherby, K., 1927- . II. Pal, S. B., 1928- . [DNLM:
1. Steroids--secretion. 2.Hormones--secretion. 3. Enzymes--physiology.
4. Endocrine Diseases. WK 150 S8355]
QP572.S7S74 1984 616.4 84-17034
ISBN 3-11-009556-4

 Printing: Gerike GmbH, Berlin. - Binding: Dieter Mikolai, Berlin. Printed in Germany.

PREFACE

Hyper- and hypo- secretion of the steroid hormones, in the absence of medication, usually denotes some underlying pathological process which may be evident at the clinical level. These modifications in secretion usually reflect changes in enzymic activity within the endocrine gland. To improve our understanding of these changes we need to increase our knowledge of the biochemical properties of the steroid converting enzymes both in the endocrine organs and, particularly, in other tissues and of the process of steroidogenesis under normal and abnormal conditions. In many instances these changes involve either a partial or complete block of a single enzymic activity with a consequent increased metabolism of the hormone precursor by other pathways which may normally be of little importance. In other instances the processes by which the changes in secretion are produced may be very much more complex. We hope that this monograph will give an up-to-date insight into the complex relationship between the steroid metabolising enzymes and disease.

K. Fotherby S. B. Pal

June 1984

CONTRIBUTORS

Numbers in parentheses indicate the page on which the authors' articles begin

Y. J. Abul-Hajj, Department of Pharmaceutical Cell Biology, College of Pharmacy, University of Minnesota, Minneapolis, Minnesota, U.S.A. (147).

R. Ducharme, Paediatric Research Centre, Hôpital Sainte-Justine and Departments of Obstetrics and Gynaecology and Paediatrics, University of Montreal, 3175 Chemin Sainte-Catherine, Montreal, Quebec H3T 1C5, Canada (175).

Bo Dupont, Human Immunogenetics Section, Memorial Sloan-Kettering Cancer Center, 1275 York Avenue, New York, New York 10021, U.S.A. (1).

W. Gibb, Paediatric Research Centre, Hôpital Sainte-Justine and Departments of Obstetrics and Gynaecology and Paediatrics, University of Montreal, 3175 Chemin Sainte-Catherine, Montreal, Quebec H3T 1C5, Canada (175).

Leonore S. Levine, Division of Pediatric Endocrinology, Department of Pediatrics, The New York Hospital-Cornell Medical Center, 525 East 68th Street, New York, New York 10021, U.S.A. (1).

V. B. Mahesh, Department of Endocrinology, School of Medicine, Medical College of Georgia, Augusta, Georgia 30912, U.S.A. (97).

Maria I. New, Division of Pediatric Endocrinology, Department of Pediatrics, The New York Hospital-Cornell Medical Center, 525 East 68th Street, New York, New York 10021, U.S.A. (1).

Songya Pang, Division of Pediatric Endocrinology, Department of Pediatrics, The New York Hospital-Cornell Medical Center, 525 East 68th Street, New York, New York 10021, U.S.A. (1).

Marilyn S. Pollack, Human Immunogenetics Section, Memorial Sloan-Kettering Cancer Center, 1275 York Avenue, New York, New York 10021, U.S.A. (1).

J. P. Preslock, Department of Obstetrics, Gynecology and Reproductive Sciences, University of Texas Medical School, Houston, Texas 77025, U.S.A. (73).

P. Saenger, Department of Pediatrics, Montefiore Hospital and Medical Center, 111 East 210th Street, Bronx, New York 10467, U.S.A. (215).

CONTENTS

ENZYMATIC DEFECTS OF ADRENAL STEROIDOGENESIS

Maria I. New*, Bo Dupont+, Songya Pang*, Marilyn S. Pollack+,
Leonore S. Levine*
*Division of Pediatric Endocrinology, Department of Pediatrics, The New York Hospital-Cornell Medical Center, 525 East 68th Street, New York, New York 10021, and
+Human Immunogenetics Section, Memorial Sloan-Kettering Cancer Center, 1275 York Avenue, New York, New York 10021, U.S.A.

Introduction

Congenital adrenal hyperplasia (CAH) is a family of inherited disorders of adrenal steroidogenesis. The disorders each have a characteristic pattern of hormonal abnormalities caused by an inherited deficiency of one of the several enzymes necessary for normal steroid synthesis.

Since the earliest documented case of virilization due to CAH was reported by the Neapolitan anatomist DeCrecchio in 1865 (1), numerous investigators have unravelled the mechanisms of adrenal steroid synthesis and the associated enzyme defects responsible for the syndromes of CAH.

I. Steroidogenesis and Enzymatic Conversions of Adrenal Steroid Hormones

A. Steroidogenesis

A simplified scheme of adrenal steroidogenesis is shown in Fig. 1. The three main classes of hormones synthesized by the adrenal cortex are mineralocorticoids (17-deoxy pathway), glucocorticoids (17-hydroxy pathway) and sex steroids. The con-

Steroid Converting Enzymes and Diseases

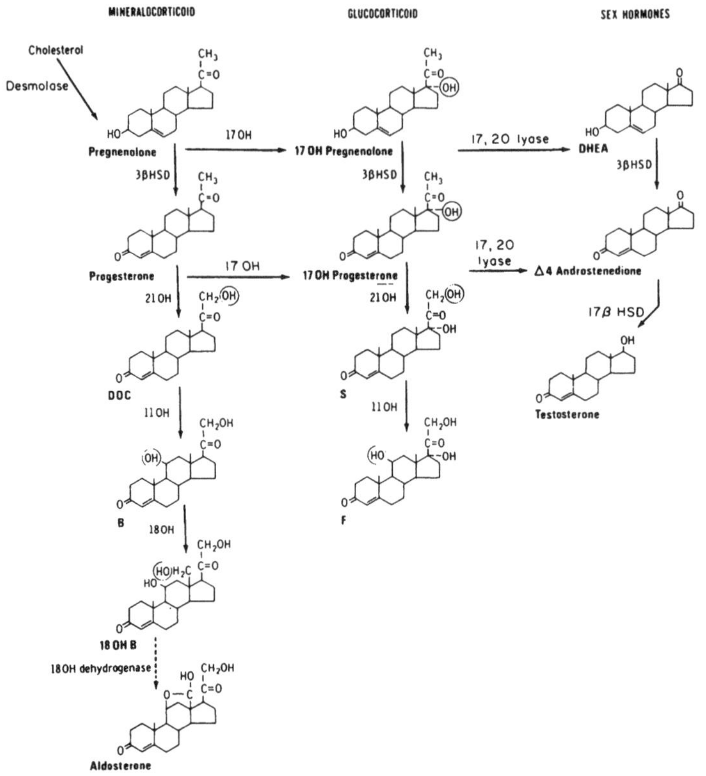

Fig. 1 Simplified scheme for adrenal steroidogenesis. Each hydroxylation step is indicated; the newly added hydroxyl group is circled. (From New and Levine (22), by permission of Plenum Press).

version of cholesterol to pregnenolone, the principal substrate for all these pathways, is increased by the binding of ACTH to receptors on the external membranes of adrenal cortical cells. This action of ACTH stimulates increased synthesis of adenosine-3',5'-monophosphate (cyclic AMP), which activates adrenal cellular phosphoprotein kinases, which in turn, catalyze side-chain cleavage, converting cholesterol to pregnenolone. Steroidogenesis has been extensively reviewed in other texts

(2,3), for the purpose of understanding the clinical manifestations of CAH, reference to the simplified scheme in Fig. 1 will suffice.

1. Mineralocorticoids (17-deoxy pathway). Pregnenolone, a Δ^5-steroid, is converted to a biologically active Δ^4-steroid, progesterone, by the enzymes 3β-hydroxysteroid dehydrogenase (3β-HSD) and an isomerase. The progesterone is then hydroxylated at C-21 to form deoxycorticosterone (DOC), an active salt-retaining hormone. When DOC is hydroxylated at C-11, corticosterone (B) is formed, which is a weak mineralocorticoid. Corticosterone is the precursor of aldosterone, the most potent salt-retaining hormone (4). Synthesis of aldosterone, a unique hormone because of the aldehyde group at C-18, occurs in the zona glomerulosa of the adrenal cortex, while synthesis of all other adrenal corticosteroids occurs in the zona fasciculata and the zona reticularis. Discussion below indicates that regulation of the zona glomerulosa differs from that of the other zones. The synthesis of aldosterone from corticosterone has not been entirely elucidated.

2. Glucocorticoids (17-hydroxy pathway). Glucocorticoid synthesis requires hydroxylation at C-17. Pregnenolone and progesterone yield 17-OH pregnenolone and 17-OH progesterone, respectively. The Δ^5-steroid 17-OH pregnenolone is converted to 17-OH progesterone, a Δ^4-steroid, by enzymatic steps similar to those that convert pregnenolone to progesterone. When 17-OH progesterone undergoes 21-hydroxylation, 11-deoxycortisol is formed, and this is further hydroxylated at C-11 to form cortisol, the most potent glucocorticoid in humans. Parallel hydroxylation steps (Fig. 1) at C-21 and C-11 of progesterone and 17-OH progesterone result in corticosterone and cortisol, respectively. Although these steps are parallel, it has not been proven that the enzymes are identical for both the 17-hydroxy and the 17-deoxy substrates.

3. Sex steroids. Dehydroepiandrosterone (DHEA) is the main unconjugated C-19 steroid secreted by the adrenal cortex. It results from the side-chain cleavage of 17-OH pregnenolone, by the action of a desmolase enzyme. DHEA, a Δ^5-steroid with little androgenic activity, is converted to Δ^4-androstenedione, a moderately active androgen, by 3β-HSD and an isomerase enzyme. When Δ^4-androstenedione is reduced at C-17, testosterone, the most potent secreted androgen in humans, is formed.

B. Mechanism of adrenal steroid regulation.

The circulating level of plasma cortisol mediates the hypothalamic-pituitary-adrenal feedback control system. The central nervous system controls the secretion of ACTH, its diurnal variation, and its increase in stress via corticotrophin-releasing factor (5, 6). Any condition which decreases cortisol secretion will result in increased ACTH secretion. In the forms of CAH in which an enzyme deficiency causes impaired cortisol synthesis, there is excessive ACTH secretion and hyperplasia of the adrenal cortex (Fig. 2).

Aldosterone secretion is primarily regulated via the renin-angiotensin system, which is responsive to the state of electrolyte balance and plasma volume (Fig. 3). Aldosterone secretion is also stimulated directly by high serum K^+ concentration. The enzyme renin, which arises from the renal juxtaglomerular apparatus, reacts with an α2-globulin produced in the liver to release angiotensin I. Angiotensin I is then enzymatically converted to angiotensin II, a potent stimulator of aldosterone secretion. This proposed mechanism is generally accepted, but there are many aspects of the regulation of aldosterone secretion which remain unexplained (7, 8).

It has been proposed by New and Seaman (9) that the zona glomerulosa and the zona fasciculata behave as two separate glands with respect to regulation and secretion. According to this concept, steroidogenesis in the fasciculata is regulated by ACTH, while that of the glomerulosa is regulated by the

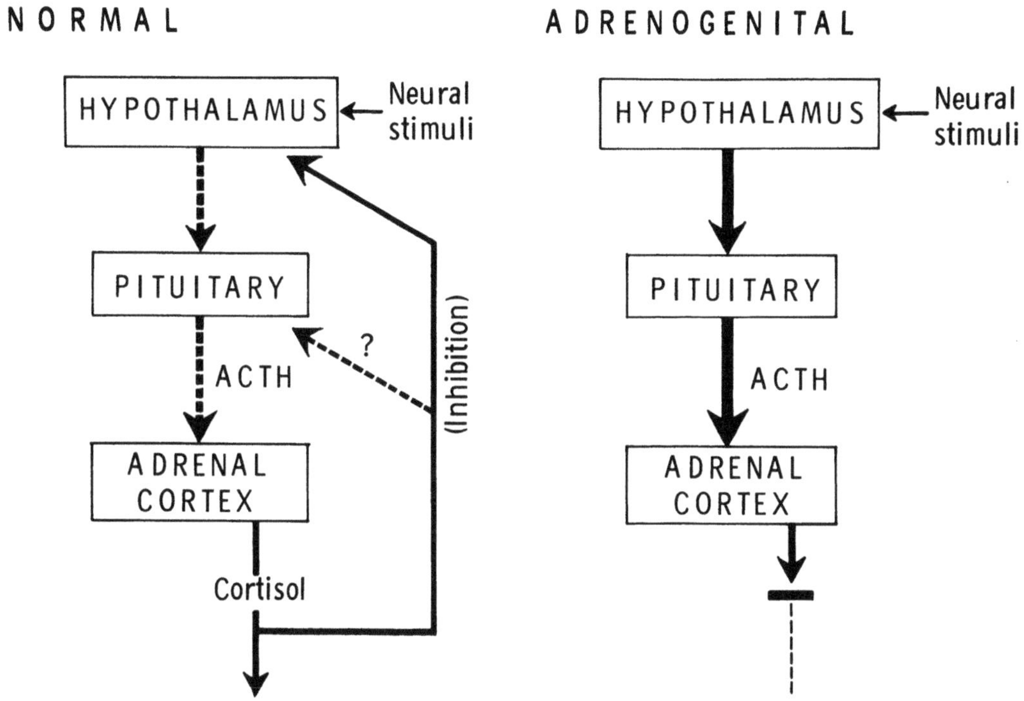

Fig. 2 The regulation of cortisol secretion in normal subjects and in patients with congenital adrenal hyperplasia. (From New and Levine (22), by permission of Plenum Press).

renin-angiotensin system such that i) ACTH stimulates secretion of cortisol, corticosterone and androgens by the zona fasciculata and zona reticularis; and ii) angiotensin stimulates aldosterone secretion by the zona glomerulosa, with ACTH exerting only a secondary influence on the glomerular secretion of aldosterone (10-12) (Fig. 4). Whereas the zona fasciculata lacks the enzyme necessary for the terminal step of aldosterone synthesis, the zona glomerulosa lacks 17α-hydroxylase activity required for the production of 17-hydroxy corticoids and androgens.

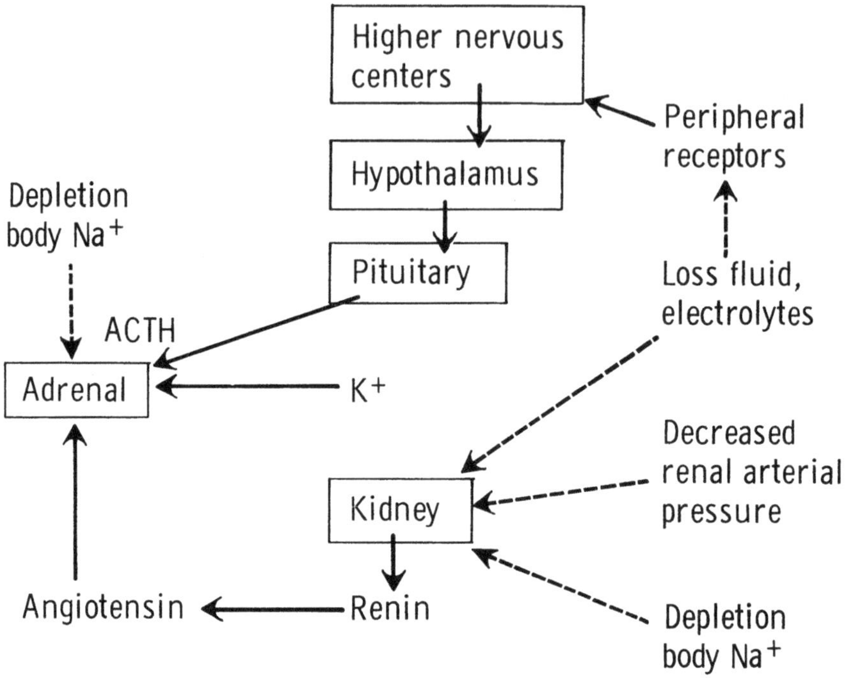

Fig. 3 Regulation of aldosterone secretion. (From New and Peterson (151), by permission).

II. Fetal Sexual Development

In order to understand the pathophysiology of CAH, it is necessary to discuss normal sexual differentiation.

According to the hypothesis developed by Jost (13), normal differentiation of male genitalia is dependent on two functions of the fetal testes:

i) Secretion of the androgen testosterone, which stimulates the Wolffian ducts to develop into the male internal genitalia (the epididymis, vas deferens, seminal vesicles and ejaculatory ducts). The secreted testosterone is also reduced in

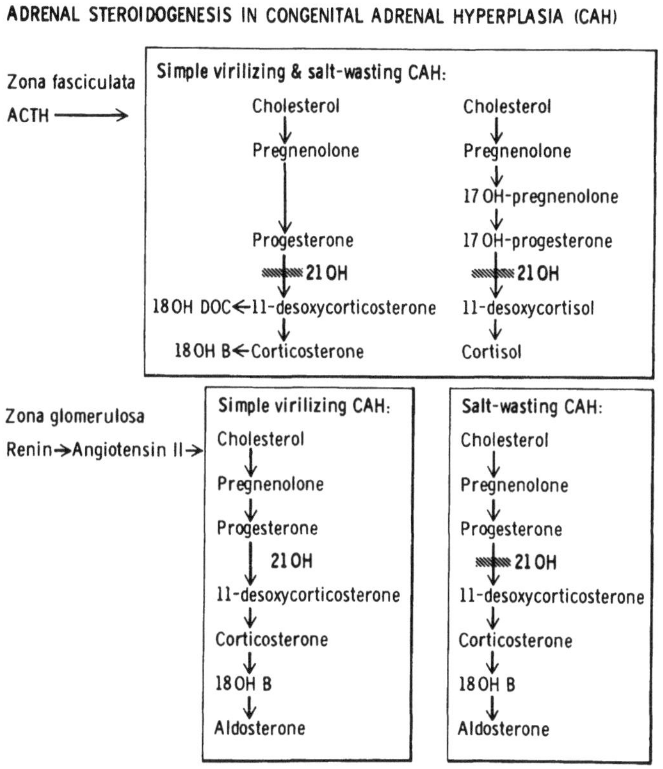

Fig. 4 Pathway of adrenal steroidogenesis in the simple virilizing and salt-wasting forms of congenital adrenal hyperplasia due to 21-hydroxylase deficiency. (From Kuhnle et al. (55), by permission of The Endocrine Society).

the target tissue to dihydrotestosterone, which causes differentiation of male external genitalia. When differentiation in the male is complete, the urethra opens at the tip of the penis. In the male, the normal source of androgen is the fetal testis, but androgen from any source can cause masculinization of external genitalia, e.g. adrenal androgen secreted during the first trimester of pregnancy.

ii) Secretion of a non-steroidal substance (14), which inhibits Mullerian duct development so that normal males are born without a uterus.

Since the fetal ovary secretes neither testosterone, nor the inhibiting factor necessary to inhibit Mullerian structures, the normal female is born without male differentiation of external genitalia (i.e. with female external genitalia), and without Mullerian repression (i.e. with a uterus and fallopian tubes). This process is shown schematically in Fig. 5.

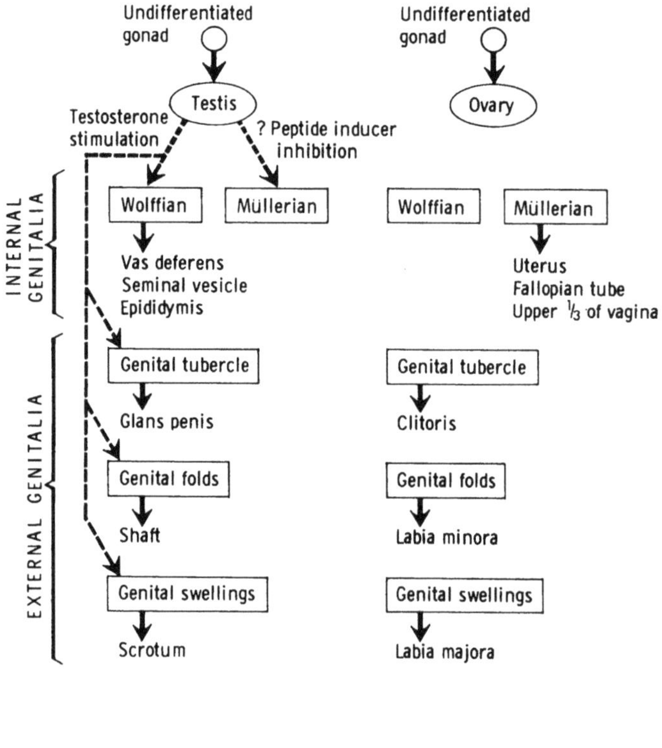

Fig. 5 Fetal sex differentiation. (From New and Levine (22), by permission of Plenum Press).

Thus, in the above scheme, the ovary does not play a determining role in sex differentiation. Female fetuses exposed to high levels of androgen, consequent to CAH, an androgen producing tumor in the mother, or administration of androgens to the mother, manifest virilization of the external genitalia, but normal internal female genitalia (15).

III. Enzyme Defects in Congenital Adrenal Hyperplasia

The following enzymatic deficiencies of steroidogenesis have been described:

21-hydroxylase — i. simple virilizing; ii. salt-wasting
11β-hydroxylase
3β-hydroxysteroid dehydrogenase (3β-HSD)
17α-hydroxylase
Cholesterol desmolase
18-hydroxylase (corticosterone methyl oxidase, Type I)
18-dehydrogenase of 18-hydroxycorticosterone (corticosterone methyl oxidase, Type II)
*17β-hydroxysteroid dehydrogenase (17β-HSD)
17-20-lyase

* may occur only in the gonad (16)

A. 21-Hydroxylase deficiency

Impairment of 21-hydroxylation is the most common enzymatic deficiency observed in CAH and results in decreased cortisol synthesis. This induces increased ACTH secretion, which has been demonstrated in the blood (11, 17, 18) and, consequently, there is overproduction of cortisol precursors and sex steroids (e.g. testosterone), which do not require 21-hydroxylase for their biosynthesis.

Early studies showed that in patients with 21-hydroxylase

deficiency, there is excessive urinary excretion of pregnanetriol, the metabolite of 17-OH progesterone (2, 19). Urinary 17-ketosteroids, which result from the metabolism of DHEA, Δ^4-androstenedione, and testosterone, are also present in increased amounts. Reliable radioimmunoassays for circulating serum levels of adrenal steroids have provided a more accurate laboratory method for the diagnosis of CAH than can be provided by urinary steroid measurement alone.

1. Simple virilizing 21-hydroxylase deficiency. The most prominent feature of this form of CAH is virilization. Since adrenocortical function begins in the third month of gestation, the fetus is exposed at the critical time of sexual differentiation to an overproduction of fetal adrenal androgens resulting from 21-hydroxylase deficiency. The excessive androgens cause virilization of external genitalia, which depend on androgens for male differentiation. Thus, in the female fetus, the external genitalia are masculinized by excessive fetal adrenal androgens resulting in female pseudohermaphroditism. In rare cases, the masculinization may be so profound that the urethra is penile (20). Since the female fetus does not have testes, Mullerian inhibiting factor is not produced, and the female with CAH is born with a uterus and fallopian tubes.

The female genital abnormalities are present only in the androgen-responsive external genitalia; internal genitalia are normal. The absence of the Wolffian system, despite increased androgen, suggests that the level of androgen necessary for Wolffian development is higher than that produced in CAH. Because CAH due to 21-hydroxylase deficiency is the most common cause of ambiguous genitalia in the newborn and because the female pseudohermaphrodite with this disorder has the capacity for an entirely normal female sex role, including fertility, it is very important to consider this disorder in any intersex problem presenting at birth.

Males with this disorder do not manifest genital abnormalities at birth. Without treatment, both males and females manifest progressive virilization, resulting in early fusion of the epiphyses, and eventual short stature.

2. Salt-wasting 21-hydroxylase deficiency. A salt-wasting form of CAH is also associated with 21-hydroxylase deficiency. The feature of virilism is the same as in simple virilizing CAH, but in the salt-wasting variety there is profound aldosterone deficiency, which is present even when the patient is sodium depleted. In the simple virilizing form there is no deficiency of aldosterone secretion. The salt-wasting form of 21-hydroxylase deficiency may present with a life-threatening crisis within the first two weeks of life, and therefore must be recognized soon after birth in order for treatment to be administered.

B. 11β-Hydroxylase deficiency

The "hypertensive" form of CAH results from a defect in 11β-hydroxylation resulting in decreased cortisol synthesis (Fig. 1). As in simple virilizing CAH, the decreased cortisol synthesis induces increased ACTH secretion (21), with the resultant overproduction of precursors.

The most prominent clinical feature of 11β-hydroxylase deficiency is also virilization. As in the simple virilizing form of 21-hydroxylase deficiency, the external genitalia of the female fetus are masculinized by excessive fetal adrenal androgen and female pseudohermaphroditism results. The internal female genitalia are normal. Postnatally, the excessive androgen production results in rapid somatic growth, advanced epiphyseal maturation, progressive penile or clitoral enlargement, early appearance of facial, axillary and pubic hair and acne. Without treatment, early epiphyseal closure, and eventual short stature result, as in 21-hydroxylase deficiency.

An additional finding in many, but not all, patients with this form of CAH is hypertension. Hypertension and lack of salt-wasting symptoms are thought to result from the excessive production of DOC, a weak mineralocorticoid. In addition to elevated serum and urinary DOC concentrations, elevated serum and urinary androgens are found. Serum levels of 11-deoxycortisol and urinary excretion of its metabolites, including the 17-hydroxysteroids, are elevated in this form of CAH (22, 23). Markedly increased serum levels and secretion rates of 11-deoxycortisol and DOC, which increase with ACTH administration, and suppressed plasma renin activity (PRA), distinguish the 11β-hydroxylase defect from 21-hydroxylase deficiency (9, 23-25).

C. 3β-Hydroxysteroid dehydrogenase deficiency

Bongiovanni (26) first described the enzymatic deficiency of 3β-HSD, which results in decreased synthesis of all three classes of adrenal steroids. There is impaired secretion of aldosterone, cortisol, and testosterone, which leads to male pseudohermaphroditism and life-threatening salt-wasting crisis in infancy. The female with 3β-HSD deficiency shows minimal clitoral enlargement, probably due to the extremely high levels of DHEA, a weak androgen. DHEA is not a sufficiently potent androgen to completely masculinize the external genitalia in the male, despite the very high amounts excreted. However, some males have been almost completely masculinized (27) by DHEA. Salt-wasting symptoms secondary to aldosterone deficiency are associated with 3β-HSD deficiency.

The urine of patients with 3β-HSD deficiency predominantly contains Δ^5 steroids. Excessive urinary excretion of DHEA and its metabolite, pregnenetriol, with markedly elevated DHEA levels in plasma, are diagnostic of 3β-HSD deficiency.

The 3β-HSD enzyme may be under different genetic control in testis, adrenal, and liver. This, in turn, could result in varying degrees of enzymatic deficiency in different organs. For instance, the testis could be less affected than the adrenal, which may explain the advent of male puberty which has been described in some cases of 3β-HSD deficiency (27, 28).

D. 17α-Hydroxylase deficiency

A defect in 17α-hydroxylase results in diminished secretion of all glucocorticoids and sex steroids, and increased secretion of mineralocorticoids. There have been additional cases reported (29, 30) since the first description of a female with 17α-hydroxylase deficiency by Biglieri et al. (31). In all cases, there was an overproduction of 17-deoxy steroids, especially corticosterone. Excessive DOC secretion is believed to be the cause of the hypertension and hypokalemia observed.

In many of the reported cases, aldosterone secretion has been low, corresponding to the low renin secretion. We have attempted to explain this low aldosterone secretion by considering the adrenal fasciculata and glomerulosa as two separate glands (29). According to our proposal, the fasciculata, which suffers from the 17α-hydroxylase defect, produces excessive DOC, which is released into the circulation. The DOC causes excessive renal tubular reabsorption of sodium which in turn, suppresses renal renin production. The glomerulosa does not receive the required renin stimulation for aldosterone synthesis, and low aldosterone secretion results. Upon administration of dexamethasone, ACTH stimulation of the fasciculata is suppressed and DOC secretion diminishes. Sodium reabsorption diminishes, permitting renin stimulation of the glomerulosa to secrete aldosterone. The observed increase in aldosterone excretion with dexamethasone makes unlikely a second enzyme defect of 18-hydroxylase, as has been proposed by others (31, 32). The corticosterone excess prevents an expression of adrenal insufficiency, even with the stress of anesthesia.

Untreated females with 17α-hydroxylase deficiency have manifested sexual infantilism. A deficiency of 17α-hydroxylation in males has resulted in male pseudohermaphroditism; internal genitalia reveal an absence of uterus and fallopian tubes, indicating the normal production of Mullerian inhibiting factor. Laboratory tests show decreased urinary excretion of 17-ketosteroids and 17-hydroxycorticoids, as well as decreased serum androgen levels.

E. Cholesterol desmolase deficiency

In 1955, Prader and Gurtner (33) described a male pseudohermaphrodite with severe salt-wasting and impaired synthesis of all three classes of adrenal corticosteroids: mineralocorticoids, glucocorticoids and sex steroids. This genetic male had female external genitalia, but male genital ducts were found at autopsy. The enlarged adrenals were remarkable in that the cortical cells were filled with lipoid material consisting of cholesterol and cholesterol esters, which prompted the name lipoid adrenal hyperplasia.

Several patients with this biosynthetic defect have been described (34, 35). Most did not survive beyond infancy, and died in adrenal crisis because of deficient mineralocorticoid and glucocorticoid production. The biosynthetic defect involves side-chain cleavage of cholesterol. At least three enzymes are involved in this conversion: 20α-hydroxylase, 22R-hydroxylase and 20α-22R desmolase (36). A deficiency of any of these three enzymes leads to profound adrenal insufficiency. Degenhart (37) produced biochemical evidence that the deficient enzyme in his patient was the cholesterol 20α-hydroxylase. The severe ambiguity of the external genitalia in the male suggests that the defect is also present in the testis. Affected female infants have normal genital development. Since the defect impairs aldosterone and DOC secretion, salt-wasting is attributable to the lack of mineralocorticoid secretion. Early recognition and proper treatment should permit survival of these infants.

F. 18-Hydroxylase deficiency

Rare cases have been described of children with 18-hydroxylase deficiency, which results in a syndrome of profound salt-wasting (37, 38). Dehydration, poor weight gain and intermittent fever are early presenting symptoms, which become less marked with age. Initially, there is hyponatremia and hyperkalemia. The penultimate step in aldosterone synthesis, conversion of corticosterone to 18-OH corticosterone, has been shown to be defective. In this disorder, there is no impairment of cortisol or sex steroid synthesis; only aldosterone and 18-OH corticosterone are deficient. Growth and development are normal and, as expected, there are no genital abnormalities or adrenal hyperplasia.

G. Deficiency of 18-dehydrogenase of 18-OH corticosterone

An adrenal glomerulosa defect in the enzymatic conversion of 18-OH corticosterone (18-OH-B) to aldosterone, causing deficient aldosterone production and resultant salt-wasting symptoms, was first described by Ulick et al. (39); numerous other cases have subsequently been reported (40, 41). A particularly high incidence of the disorder has been found in an inbred population of Iranian Jews (42). The glucocorticoid and androgen pathways are spared in this enzyme deficiency, there is no hypersecretion of ACTH or genital abnormalities. The clinical severity ranges from salt-wasting crises leading to infant death, to intermediate cases of short stature and postural hypotension and to cases of asymptomatic adults whose deficiency was only detectable by biochemical tests. Variable levels of aldosterone secretion have been reported, indicating a variability in the severity of the enzyme defect. In some instances, normal or even elevated plasma aldosterone concentrations have been found in salt-depleted patients, leading to confusion in the diagnosis of the disorder. However, these aldosterone levels always exist in the presence of markedly elevated plasma renin activity. Thus, although the aldosterone response might appear to fall within a "normal" range in absolute terms, it is inadequate in relation to the degree of the stimulus of sodium depletion, indicated by the substantially elevated plasma renin activity. In general, the best diagnostic measure for specifying the enzymatic defect is to examine the precursor-to-product ratio, when the trophic hormone is increased. The diagnosis of

18-dehydrogenase deficiency can be made by comparing the secretion of aldosterone to that of 18-OH-B, measured by their urinary metabolites. Thus, the ratio of 18-OH-B metabolite to aldosterone, normally less than 3.0, is often greater than 100 in untreated patients, suggesting that the zona glomerulosa cannot effectively convert 18-OH-B to aldosterone (42). The term "corticosterone methyl oxidase, Type II deficiency" has recently been adopted as a more accurate name for this enzyme disorder, and "corticosterone methyl oxidase, Type I deficiency" for 18-hydroxylase deficiency (40).

H. 17β-Hydroxysteroid dehydrogenase (17β-HSD) deficiency

Deficiency of the 17β-HSD enzyme also results in decreased testosterone synthesis and is another known cause of male pseudohermaphroditism (16, 43-45). With one exception (46), all previously reported genetic males were assigned to the female sex of rearing at birth, because they were so incompletely masculinized. The conversion of Δ^4-androstenedione (Δ^4) to testosterone is impaired in the testis, although peripheral conversion of Δ^4 to testosterone is apparently still intact. This suggests that the enzymes controlling 17β-HSD in the testis and in the liver are under different genetic control. Elevated Δ^4/testosterone ratios in peripheral and spermatic blood of older patients are diagnostic of the 17β-HSD deficiency. In newborns and infants, hCG stimulation may be required to unmask the abnormal Δ^4/testosterone ratios (47). Conversion of estrone to estradiol is also impaired in 17β-HSD deficiency, but usually to a lesser degree. The defect is probably only expressed in the gonads and not in the adrenal.

I. 17-20-Lyase deficiency

The 17-20-lyase enzyme deficiency results in male pseudohermaphroditism due to a partial defect in the conversion of 17-OH progesterone and 17-OH pregnenolone to C_{19} steroids by the testis and the adrenal; only sex steroid biosynthesis is affected. Urinary pregnanetriolone, one of the metabolites of 17-OH pregnenolone, is increased, and ACTH and hCG stimulation increase it further. However, neither testosterone nor DHEA excretion rises appreciably. Seven patients in a total of three different kindred with this disorder have been reported; all seven had XY karyotypes (48-50).

Table 1 summarizes the clinical and laboratory features of these enzyme disorders.

IV. The Zona Fasciculata and Zona Glomerulosa as Separate Glands

In concert with the theory that the zona fasciculata and zona glomerulosa function as two separate glands under different regulatory control, New and Seaman (9) proposed that it was possible for an enzyme deficiency to occur in the fasciculata, but not in the glomerulosa. Several reports have suggested that this is the case for the genetic defect of 11β-hydroxylation (51-54). Recent extensive studies in four patients with 11β-hydroxylase deficiency have confirmed that patients can at once have defective 11β-hydroxylation in the fasciculata, and normal 11β-hydroxylation in the glomerulosa (23).

It is also proposed that the independence of enzyme activity between the zona fasciculata and the zona glomerulosa may account for the different degrees of severity in the two different clinical forms of 21-hydroxylase deficiency: the simple virilizing form and the salt-wasting form (55).

The hypotheses traditionally proposed to explain the difference between the salt-wasting and simple virilizing forms of CAH are based on either a "one-enzyme" or a "two-enzyme" defect. However, neither *in vivo*, nor *in vitro* studies have been conclusive, and results compatible with both theories have been reported (56).

The "one-enzyme" theory attributes the differences between salt-wasting and simple virilizing CAH to a different degree of

Table 1. Clinical and Laboratory Features of Various Disorders of Adrenal Steroidogenesis

Clinical Features						Laboratory Findings								
Newborn with sexual ambiguity						Urinary Excretion				Circulating Hormones				
Female	Male	Salt wasting	Hyper-tension	Postnatal virilization	Enzyme deficiency	17-KS	17-OH	P'triol	Aldo	17-OHP	Δ4	DHEA	Testosterone	Renin
+	0	0	0	+	21-Hydroxylase: -simple virilizing	↑↑	nl or ↓	↑↑	nl	↑↑	↑↑	nl or ↑ (DHEA/Δ4 ↓)	↑	nl or ↑
+	0	+	0	+	-salt-wasting	↑↑	↓	↑↑	↓	↑↑	↑↑	nl or ↑	↑	↑↑
0	0	0	0	+	-"late onset"	↑	nl	↑	nl	↑	↑	nl or ↑	nl or ↑	nl
+	0	0	+	+	11β-Hydroxylase	↑↑	↑↑ [a]	↑	↓	↑	↑↑	↑	↑	↓↓
+	+	+	0	0[f]	3β-HSD[b]	↑ [c]	↓↓	nl or ↓	↓	nl or ↑	nl or ↑	↑↑↑	d	↑
0	+	0	+	0	17α-Hydroxylase	↓↓	↓↓	↓↓	↓	↓	↓	↓	↓	↓
0	+	+	0	0	Cholesterol desmolase	↓↓	↓↓	↓↓	↓↓	↓	↓	↓	↓	↑
0	0	+	0	0	18-Hydroxylase	nl	nl	nl	↓↓	nl	nl	nl	nl	↑
0	0	+	0	0	Methyloxidase type II	nl	↑ [e]	nl	↓↓	nl	nl	nl	nl	↑
?	+	-	-	+[f]	17β-HSD[g]	nl or ↑	nl	nl	nl	nl	↑↑	nl or ↑	nl or ↓ (Δ4/T ↑↑)	nl
?	+	-	-	+[f]	17,20-Lyase	↓	nl	nl or ↑	nl	nl or ↑	↓↓	nl or ↓	↓↓	nl

nl = normal

P'triol = Pregnanetriol

a Mostly THS

b The values presented apply to the infant and the very young child

c Mostly Δ5-17 ketosteroids

d ↓ or normal in male
↑ or normal in female

e Largely 18-hydroxy THA, which gives a Porter-Silber reaction

f Only in males at puberty

g This defect may occur only in the testes

enzymatic deficiency, the most severe deficiency leading to salt-wasting. The "two-enzyme" theory postulates that the simple virilizers have an enzymatic deficiency in the 17-hydroxy (glucocorticoid) pathway, whereas the salt-wasters have a deficiency in both the 17-hydroxy and 17-deoxy pathways of adrenal steroidogenesis. This theory is supported by normal, or even elevated aldosterone secretion or excretion in simple virilizers, in contrast to the impaired aldosterone production in salt-wasters (56).

Recent studies from our laboratory (55, 56) have suggested a new hypothesis to explain the difference in salt-wasting and simple virilizing CAH which conforms to the two gland model of the adrenal cortex. This hypothesis states: a) in both simple virilizers and salt-wasters, there is a fasciculata defect of 21-hydroxylation in both the 17-hydroxy and 17-deoxy pathways; and b) in the salt-waster there is a defect in 21-hydroxylation in the glomerulosa as well, while in the simple virilizer the glomerulosa is spared this defect. This hypothesis is schematically presented in Fig. 4, and the data supporting this hypothesis are presented in Figs. 6 and 7.

Kuhnle's study (55) verifies this hypothesis by demonstrating a 21-hydroxylase deficiency in the 17-deoxy pathway of the fasciculata in both simple virilizers and salt-wasters. Under ACTH stimulation, progesterone levels rose markedly, while DOC and corticosterone responses were minimal. When the fasciculata was then suppressed with dexamethasone, and the glomerulosa stimulated with low sodium intake, both normal subjects and simple virilizers, but not salt-wasters, demonstrated a rise in serum aldosterone, urinary aldosterone, and 18-OH corticosterone, indicative of normal glomerulosa 21-hydroxylase function. These results indicate the absence of a 21-hydroxylase deficiency in the zona glomerulosa of the simple virilizer, and show that in both normal subjects, and in the simple virilizers, the zona glomerulosa can increase aldosterone secretion in response to renin stimulation independent of precursor hormones of the zona fasciculata (Fig. 7).

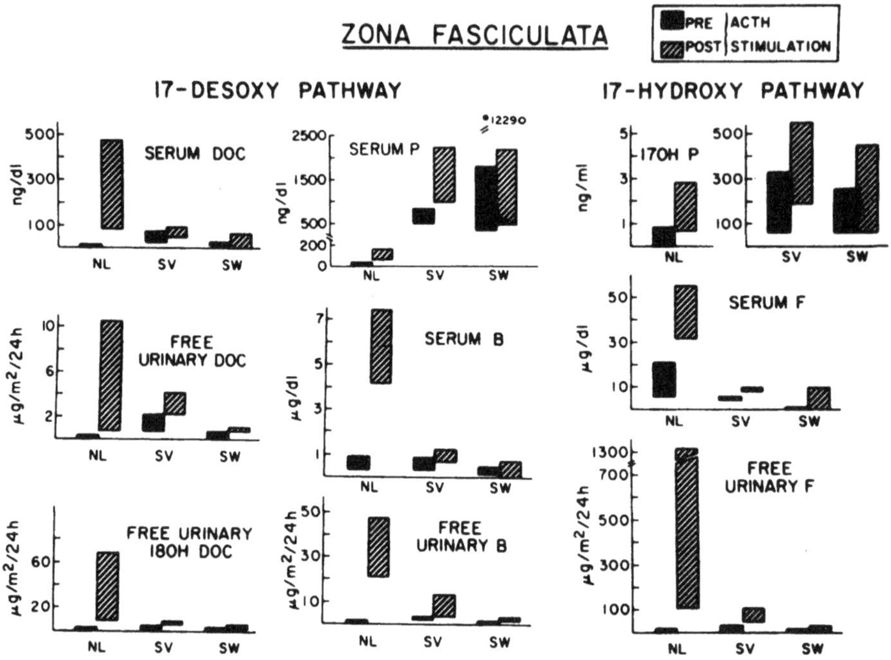

Fig. 6 Hormonal response to ACTH stimulation of the zona fasciculata. NL, normal subject; SV, patient with the simple virilizing form of 21-hydroxylase deficiency; SW, patient with the salt-wasting form of 21-hydroxylase deficiency. (From Kuhnle et al. (55), by permission of The Endocrine Society).

The findings in 21-hydroxylase deficiency are consistent with the reports of an enzymatic deficiency only in the fasciculata, sparing the glomerulosa in CAH due to 11β-hydroxylase deficiency (23, 52). The findings are also particularly noteworthy in view of our proposal that the independent functioning of the fasciculata and glomerulosa may account for the low levels of aldosterone secretion found in 17α-hydroxylase deficiency.

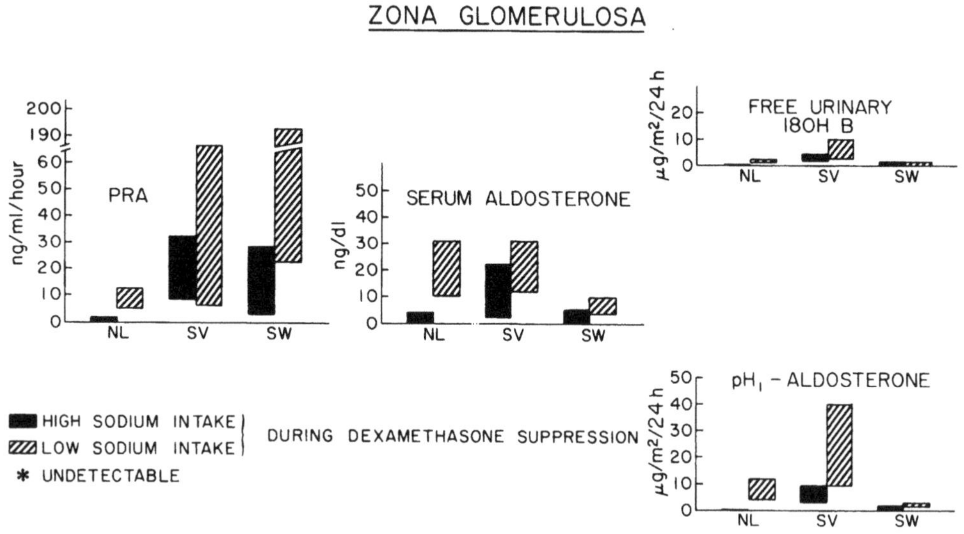

Fig. 7 Hormonal response to sodium deprivation of the zona glomerulosa. See legend to Fig. 6.

These findings further suggest the possibility of separate genetic loci for the regulation of 21-hydroxylation in the zona fasciculata and the zona glomerulosa. Genetic studies of recombination, discussed below, are also suggestive of this possibility. The hormonal data suggest that within the fasciculata, only one genetic locus seems to regulate 21-hydroxylase enzyme activity in both the 17-hydroxy and 17-deoxy pathways.

V. Treatment of Congenital Adrenal Hyperplasia

The virilization caused by CAH can be remedied by surgical correction of the ambiguous genitalia, but the decision to do so must take into account an assessment of the patient's potential for future sexual function and fertility. In cases of female pseudohermaphroditism due to 21- or 11β-hydroxylase deficiency, the aim of surgical repair should be to remove the redundant erectile tissue, preserve the sexually sensitive glans clitoris, and provide a normal vaginal orifice that will function adequately for menstruation and intromission (57). When there is early therapeutic intervention in these patients, normal puberty, fertility and childbearing are possible because of the normal internal genitalia. In contrast, in cases of male pseudohermaphroditism due to 3β-HSD or 17α-hydroxylase deficiency, a female sex assignment may be advisable due to the inability to repair the external genitalia to conform to a male sex assignment.

The aim of endocrine therapy in CAH is to provide replacement of the deficient hormones. Since 1949, when Wilkins et al. (58) and Bartter (59) discovered the efficacy of cortisone therapy for CAH due to 21-hydroxylase deficiency, glucocorticoid therapy has been the keystone of treatment for this disorder. Glucocorticoid administration both replaces the deficient cortisol and suppresses ACTH overproduction.

Plasma renin activity (PRA) has long been recognized to be elevated in the simple virilizing (non-salt-wasting) form, as well as in the salt-wasting form (56, 59), although aldosterone levels have not been reported as being deficient in the simple virilizing form of 21-hydroxylase deficiency (56). Despite the observation of elevated PRA, it had not been customary to supplement conventional glucocorticoid replacement therapy with the administration of salt-retaining steroids in cases of simple virilizing 21-hydroxylase deficiency. However, in a recent clinical study, Rosler et al. (60) demonstrated that the addition of salt-retaining hormone to glucocorticoid therapy in simple virilizing patients with elevated PRA does improve the hormonal control of the disease.

Rosler showed that in patients with CAH due to 21-hydroxylase deficiency, the PRA was closely correlated with the ACTH level (Fig. 8). Thus, when PRA was normalized by the administration of 9α-fludrohydrocortisone acetate, a steroid with salt-retaining activity, the ACTH level fell, and excessive

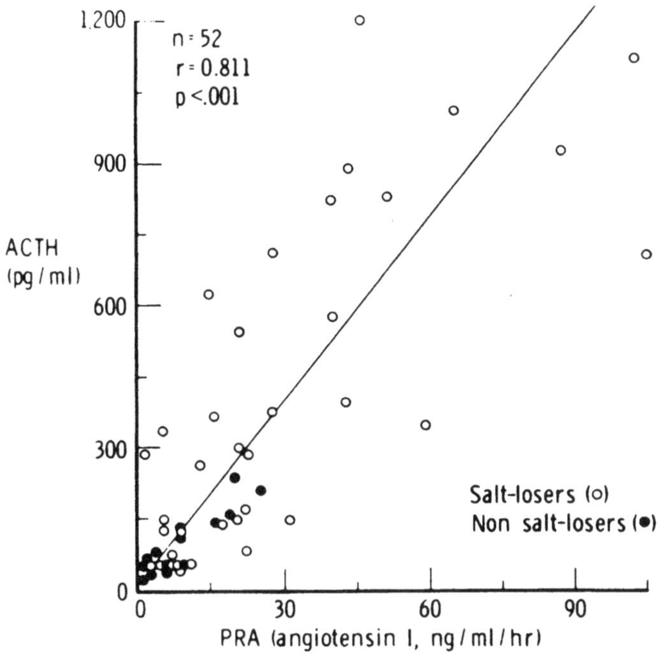

Fig. 8 Correlation between ACTH and plasma renin activity (PRA) in patients with CAH treated with constant replacement doses of glucocorticoids equivalent to 25 mg/m^2/day of hydrocortisone. Patients were studied during different states of sodium balance. (From Rosler et al. (60), by permission of The Endocrine Society).

androgen stimulation by ACTH decreased (Fig. 9). The addition of salt-retaining steroids to the therapeutic regimen often made possible a decrease in the glucocorticoid dose. Normalization of PRA also resulted in improved statural growth (56).

In the past, urinary 17-ketosteroids and pregnanetriol excretion were the biochemical monitors of hormonal control.

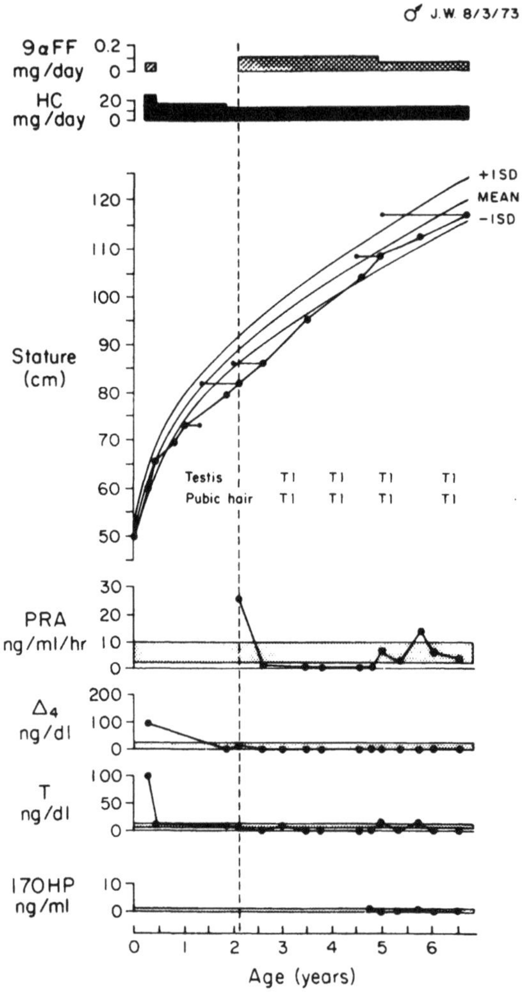

Fig. 9 Growth curve of a patient with salt-wasting congenital adrenal hyperplasia before and after therapeutic control of plasma renin activity. Note that with renin suppression, hydrocortisone dose could be lowered, growth improved, and androgen suppression was maintained despite the decrease in hydrocortisone dose. (From New et al. (56), by permission of Academic Press).

Using radioimmunoassays, it has been possible to establish normal serum androgen concentrations for children of various ages (61-80). Recent studies have indicated that serum 17-OH progesterone (17-OH-P) and Δ^4 levels provide a sensitive index of biochemical control. The serum testosterone level is useful in females and prepubertal males, but not in newborn and pubertal males (81-93).

The combined laboratory determinations of PRA, 17-OH-P and serum androgens, and the clinical assessment of growth and pubertal status, must all be considered in adjusting the dose of glucocorticoid and salt-retaining steroid. In our clinic, hydrocortisone and 9α-fludrohydrocortisone acetate are used as treatment modalities. This approach to therapy has recently been confirmed by Winter (94). Measurement of PRA can be used to monitor efficacy of treatment not only in 21-hydroxylase deficiency, but also in the other salt-losing forms of CAH (cholesterol desmolase and 3β-HSD). PRA is also useful as a therapeutic index in those forms of CAH with mineralocorticoid excess and suppressed PRA (11β-hydroxylase and 17α-hydroxylase). In poor control, PRA is elevated in the salt-losing forms and suppressed in the mineralocorticoid excess forms (Fig. 10).

VI. Genetics of Congenital Adrenal Hyperplasia

A. 21-Hydroxylase deficiency

1. Population studies. Several surveys have established that 21-hydroxylase deficiency is transmitted as an autosomal recessive trait (20, 95, 96), with males and females equally at risk (97). With few exceptions (98), the severity of the 21-hydroxylase deficiency, which parallels the degree of salt-wasting, is consistent within one family.

In Europe and the USA, recent estimates of the incidence of CAH have been between 1:5,000 and 1:15,000. In Alaska, the incidence in Yupik-speaking Eskimos is unusually high, while

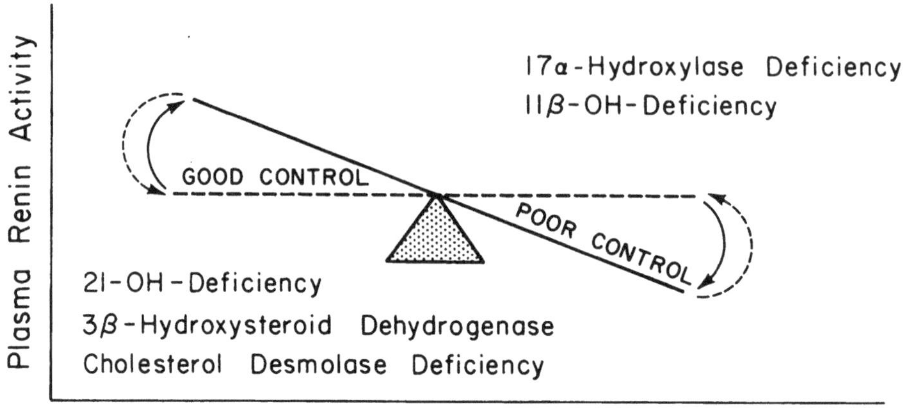

Fig. 10 The pivotal role of monitoring plasma renin activity in various forms of congenital adrenal hyperplasia. In poor control, renin is elevated in 21-hydroxylase deficiency, 3β-hydroxysteroid deficiency and cholesterol desmolase deficiency and decreases with proper mineralocorticoid treatment. In contrast, in 17α-hydroxylase deficiency and 11β-hydroxylase deficiency, renin is suppressed in poor control and rises with proper treatment. (From New et al. (56), by permission of Academic Press).

the low incidence reported in Maryland (95) may have been due to inadequate case ascertainment (Table 2). The gene frequency is estimated as approximately one in one hundred (99). The salt-wasting variety occurs in about 30-80 per cent of patients with 21-hydroxylase deficiency (100, 101). In Alaska, the Yupik Eskimos exhibit a very high incidence of the salt-wasting variety (102).

The 11β-hydroxylase, 3β-HSD, and 17α-hydroxylase deficiencies are also transmitted as autosomal recessive traits, but are rarer than the 21-hydroxylase deficiency. The mode of inheritance of the other enzyme defects has not been documented.

Table 2. Estimated Incidence of CAH (21-hydroxylase deficiency)

Reference	Population	Patient/Live Births
Childs *et al*. 1956 (95)	Maryland, U.S.A.	1:67,000
Prader 1958 (96)	Zurich, Switzerland	1:5,041
Hubble 1966 (149)	Birmingham, England	1:7,255
Rosenbloom and Smith 1966 (98)	Wisconsin, U.S.A.	1:15,000
Hirshfeld and	Alaska, U.S.A	1:700*
Fleshman 1969 (102)	Yupik Eskimo, U.S.A.	1:245
Qazi and Thompson 1972 (150)	Toronto, Canada	1:13,000*
Mauthe *et al*. 1977 (151)	Munich, Germany	1:9,831
Muller *et al*. 1979 (99)	Tyrol, Austria	1:8,991
Werder *et al*. 1980 (133)	All of Switzerland	1:15,472

* Incidence was corrected for both variants of salt-waster and non-salt-waster.

2. HLA linkage to 21-hydroxylase deficiency. The genes for HLA (human leukocyte antigens), cell surface antigens important in transplantation, are located on chromosome 6. The HLA complex consists of at least four genetic loci which code for the antigens HLA-A, HLA-B, HLA-C and HLA-D/DR. Multiple alleles have been demonstrated for each locus. In addition to the HLA loci, several other loci have been mapped on chromosome 6 in close linkage with HLA (Fig. 11). Each individual inherits one chromosome 6 from his father and one from his mother. The HLA genes

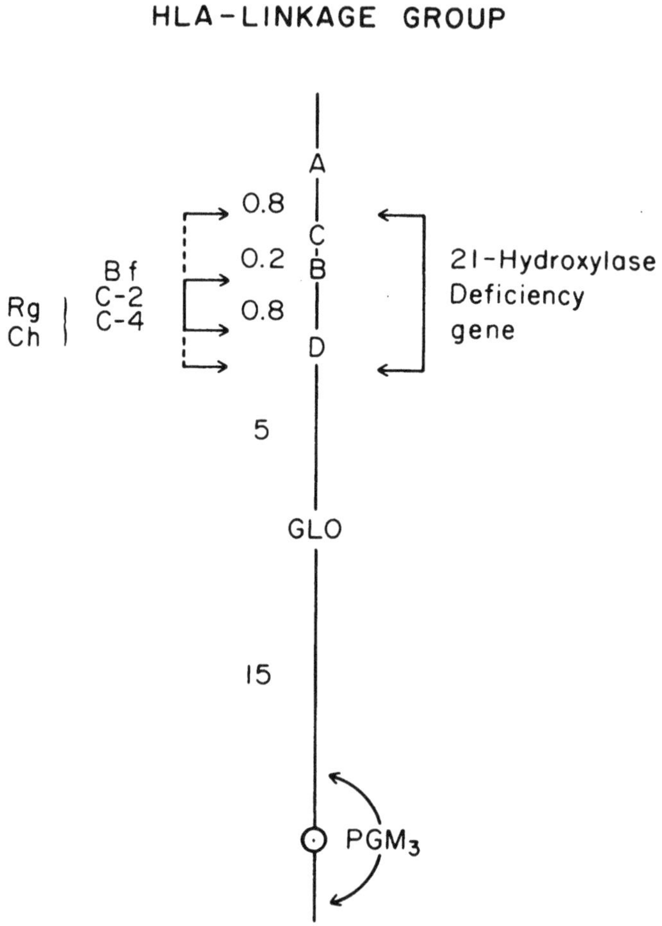

Fig.11 HLA linkage group on chromosome 6. The recombinant fractions for the known linkages between A:C, C:B, B:D, B:GLO and GLO:PGM_3 are shown. The position of the genes for factor B (Bf), complement C-2, complement C-4, (the Rodgers (Rg) and Chido (Ch) blood groups) is also indicated. The 21-hydroxylase deficiency gene can be mapped between HLA-A and Glyoxalase I (GLO). The most likely position of the 21-hydroxylase deficiency gene is very close to HLA-B. (From Levine et al. (104), by permission of The New England Journal of Medicine).

are codominantly expressed as illustrated below:

$$\frac{\text{A3, Bw47(w4), Cw6, DR7}}{\text{A28, Bw35(w6), Cw4, DR5}}$$

in which one haplotype (the set of A3, Bw47(w4), Cw6, DR7) is inherited from one parent, while the other haplotype (A28, Bw35(w6), Cw4, DR5) is inherited from the other.

Close genetic linkage between HLA and CAH due to 21-hydroxylase deficiency was first described in 1977 (103). In this initial study, HLA genotyping of parents and children in six families with one or more children affected with CAH due to 21-hydroxylase deficiency was performed. In five of these families, all the affected offspring were HLA identical, and different from their unaffected sibs. In the sixth family, the two affected children were HLA-B identical. Subsequently, studies of 34 unrelated families with a total of 48 patients were reported from New York and Zurich (104). The findings of this study confirmed the HLA linkage of 21-hydroxylase deficiency, as illustrated in the two typical pedigrees shown in Fig. 12. As can be observed in Family 7, the three affected sibs are HLA identical. In Family 16, the unaffected sibs are all HLA different from their affected sister. Each parent is an obligate heterozygote carrier and has transmitted one HLA haplotype carrying the gene for 21-hydroxylase deficiency to the patient. Since the HLA linkage is established, the brother and sister having only one haplotype linked to the gene for CAH are presumed heterozygotes. The sib sharing neither haplotype with the patient is presumed not to carry the gene for 21-hydroxylase deficiency. Thus, the HLA genotype can be a marker for the CAH genotype.

Recent international studies have provided more detailed genetic mapping of the specific location of the 21-hydroxylase deficiency gene relative to the different loci that constitute the HLA complex. These studies have established that the 21-hydroxylase deficiency gene is located between the HLA-A locus

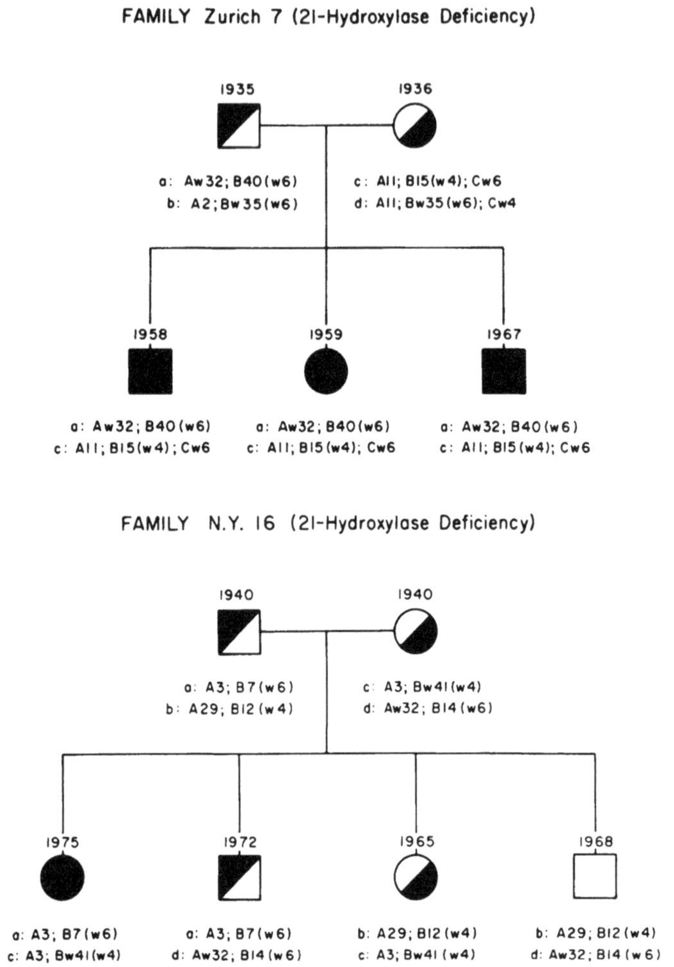

Fig. 12 Pedigrees for two families with classical 21-hydroxylase deficiency. The HLA haplotypes for the HLA-A, HLA-B and HLA-C alleles are given in each family. The paternal haplotypes are labeled a and b, and the maternal haplotypes c and d. The parents are obligate heterozygous carriers for the 21-hydroxylase deficiency gene (denoted by the half-black symbols). The affected children are denoted

by black symbols. In Family 7, three affected sibs are HLA genotypically identical. In Family 16, one affected child is HLA genotypically different from the three unaffected sibs. One sib who carries the parental a and d haplotypes is presumed to be a heterozygous carrier for 21-hydroxylase deficiency because he shares the a haplotype with the patient. Another sib has the parental b and c haplotypes and shares the c haplotype with the patient, and should be a carrier of the classical 21-hydroxylase deficiency gene. The child with the b and d haplotypes should be normal for the gene. (From Levine et al. (104), by permission of The New England Journal of Medicine).

and the centromere, but is further away from the centromere than the glyoxalase I locus (105). The gene is thus mapped to within 3-4 centimorgans.

Statistical methods of genetic analysis have demonstrated close genetic linkage between the 21-hydroxylase deficiency gene and HLA. The studies found a peak LOD Score of 15.65 at a recombination frequency of $\theta=0.00$, for linkage between HLA and 21-hydroxylase deficiency CAH (105). The LOD Score is a statistical index of the certainty with which one can arrive at a conclusion of genetic linkage. A score of 15.65 means that the odds are $10^{15.65}$ to 1 that linkage exists. In humans, genetic linkage is considered established if the LOD Score exceeds 3.00.

3. Heterozygote detection. Attempts to detect heterozygosity for the 21-hydroxylase gene by hormonal tests prior to the discovery of HLA linkage were only partially successful. In studying the obligate heterozygote parents of children with CAH, a mild deficiency of 21-hydroxylase was demonstrated in several studies (106, 107). However, similar studies in the sibs of patients with CAH were difficult to interpret because of the inability to ascertain which sibs are carriers of the gene and which are genetically unaffected. With the demonstration of linkage between the genes for HLA and 21-hydroxylase deficiency,

HLA genotyping makes it possible to predict which sibs are carriers and which sibs are genetically unaffected. The method for utilizing HLA genotyping in this prediction is shown in Fig. 12 (Family 16). The validity of the prediction of heterozygosity by HLA genotyping is supported by hormonal studies. Thus, in family studies, the response of 17-OH-P to ACTH stimulation was higher in family members predicted by HLA genotyping to be heterozygotes, than in family members predicted to be genetically unaffected (Fig. 13) (108). No other hormonal measurement was useful in discriminating heterozygotes from normals (108, 109). Other investigators observed similar correlations between HLA genotyping and hormonal measurements (110). Grosse-Wilde et al. (110) and Lejeune-Lenain et al. (111) have recently reported obtaining improved segregation of CAH heterozygotes from unaffected subjects in the adult female population, by pretreating these subjects with dexamethasone.

4. Non-classical variants of 21-hydroxylase deficiency. Symptomatic non-classical 21-hydroxylase deficiency (late onset or "acquired" adrenal hyperplasia) is a syndrome that presents in later childhood or early adulthood with virilization and menstrual disturbances associated with endocrinological features consistent with 21-hydroxylase deficiency (112-115). Unlike patients with classical 21-hydroxylase deficiency, females with symptomatic non-classical 21-hydroxylase deficiency demonstrate no evidence of *in utero* virilization and are born with normal vaginal and urethral orifices without labial fusion. Both types of patients respond similarly to glucocorticoid treatment. The late presentation of a biochemical defect has raised the question as to whether this is the same inherited disorder as classical CAH with delayed presentation, or is an acquired disorder distinct from CAH. Although initial studies suggested that the late onset disorder was not HLA-linked (116, 117), more recent reports have provided evidence that it is (118-121), and it has been proposed that classical and non-

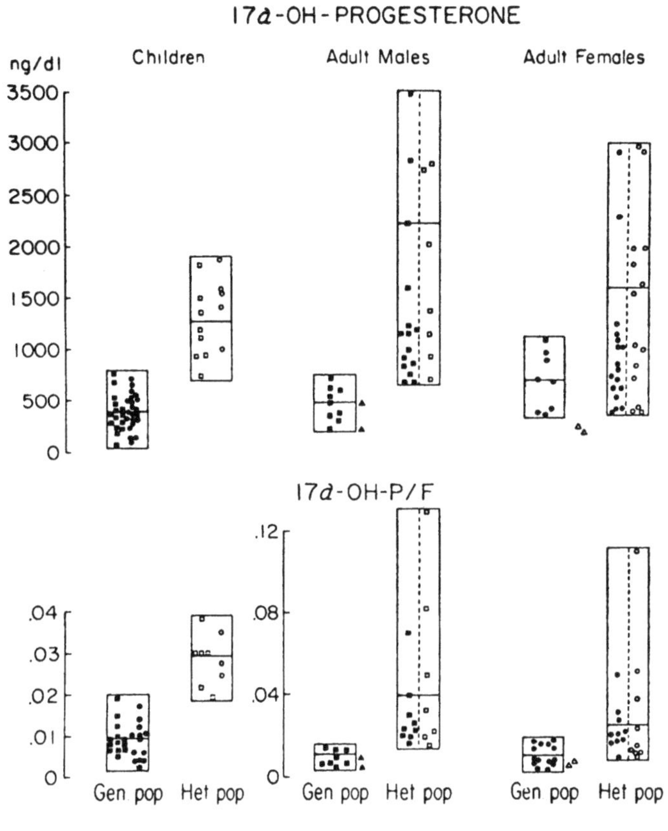

Fig. 13 Stimulated 17-OHP level and the stimulated 17-OHP to F ratio in prepubertal and early pubertal children, postpubertal males, and postmenarchal females. Gen. pop., general population;

■ indicates male

● indicates female

▲ indicates the homozygous normal brother of a patient with CAH, based upon HLA typing.

Δ indicates the homozygous normal sister of a patient with CAH, based upon HLA typing.

Het. pop., heterozygous population;

■ ● indicates heterozygous parents, based upon HLA typing.

□ o indicates heterozygous sibs or family members, based upon HLA typing.

The bar represents the range; the heavy horizontal line represents the mean (From Lorenzen et al. (108), by permission of The Endocrine Society).

classical 21-hydroxylase deficiencies are allelic variants (56, 120, 121).

Our HLA studies of families with classical 21-hydroxylase deficiency have led to a new, asymptomatic form of non-classical 21-hydroxylase deficiency. In the course of HLA genotyping and hormonal testing, we encountered family members whose biochemical profiles were characteristic of 21-hydroxylase deficiency, although the clinical hallmarks commonly accompanying the disorder (e.g. virilization, abnormal puberty and growth, infertility) were absent. The presence of hormonal abnormalities without clinical stigmata led to the designation of this disorder as "cryptic" 21-hydroxylase deficiency. The hormonal abnormalities which these asymptomatic family members demonstrated were similar to those observed in the symptomatic non-classical form of 21-hydroxylase deficiency (122). Further investigation suggested that these asymptomatic relatives of patients with 21-hydroxylase deficiency were genetic compounds, having 21-hydroxylase deficiency as a result of inheriting two recessive genetic defects: 1) a severe 21-hydroxylase genetic defect shared with the index case with classical CAH (21-OH^{SEVERE}); and 2) a mild, non-classical variant of the 21-hydroxylase genetic defect (21-OH^{MILD}). A 21-OH^{SEVERE}/21-OH^{MILD} genotype was thus assigned to these family members with non-classical 21-hydroxylase deficiency.

LOD Score analysis established close genetic linkage between HLA and the 21-OH^{MILD} gene (122). The pedigrees of several families with members having the asymptomatic disorder are illustrated in Fig. 14 and 15.

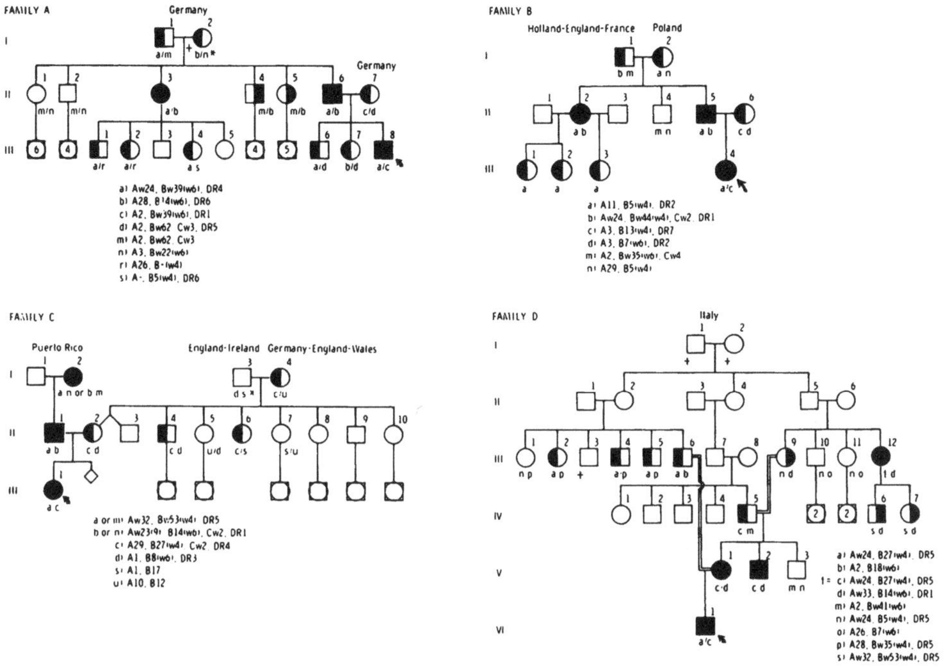

Fig. 14 Pedigrees of families A-D with classical and asymptomatic non-classical 21-hydroxylase deficiency. The HLA genotypes for HLA-A, HLA-B, HLA-C, and HLA-DR are indicated for each family member tested. The index case with CAH is assigned the haplotypes a/c; * indicates haplotypes deduced from offspring.

indicates heterozygous carriers of the severe classical deficiency gene for CAH.

indicates family members with asymptomatic non-classical 21-hydroxylase deficiency, having both a severe and a mild deficiency gene.

indicates heterozygous carriers of the mild non-classical 21-hydroxylase deficiency gene.

indicates offspring, number if known.

(Adapted from New et al. (56), by permission of Academic Press).

Zachmann and Prader (123, 124) also reported family members of patients with classical 21-hydroxylase deficiency whose hormonal responses are consistent with asymptomatic 21-hydroxylase deficiency. They proposed that these members were unusually manifesting heterozygotes for the classical gene (21-OH ^{SEVERE}/21-OH^{NORMAL}), rather than genetic compounds (21-OH ^{SEVERE}/21-OH^{MILD}) as we have proposed. Our demonstration that family members predicted by HLA genotyping to be heterozygous for the mild deficiency gene (21-OH^{MILD}/21-OH^{NORMAL}) have a 17-OH-P response to ACTH identical to that of family members predicted to be heterozygous for the classical gene (21-OH ^{SEVERE}/21-OH^{NORMAL}), argues against the suggestion of Zachmann and Prader. In Family H (Fig. 15) for example, members II-3, III-3, and III-4 all demonstrated hormonal responses characteristic of heterozygous carriers for a 21-hydroxylase deficiency gene; HLA genotyping revealed they shared one HLA haplotype ("b") with the asymptomatic family members (II-2, II-4) but no HLA haplotype with the index case with classical CAH (III-5). If, as proposed by Zachmann and Prader, the asymptomatic members II-2 and II-4 were simply unusual heterozygotes for the classical gene, members II-3, III-3 and III-4 would be predicted to be genetically unaffected (21-OH^{NORMAL}/21-OH^{NORMAL}). This prediction conflicts with the observed hormonal responses in these members, which clearly fall in the range for heterozygotes. The proposal that these members have inherited a variant 21-OH^{MILD} gene from a family member with asymptomatic non-classical 21-hydroxylase deficiency is consistent with the hormonal findings. Subsequent genetic studies have indicated that both asymptomatic and symptomatic non-classical 21-hydroxylase deficiencies may result either from a genetic compound of the classical and non-classical deficiency genes, or from homozygosity for the mild non-classical deficiency gene.

An interesting biological paradox is the absence of clinical signs in patients with asymptomatic non-classical 21-hydoxylase deficiency, in contrast with the severe virilization noted in patients with symptomatic non-classical 21-hydroxylase defi-

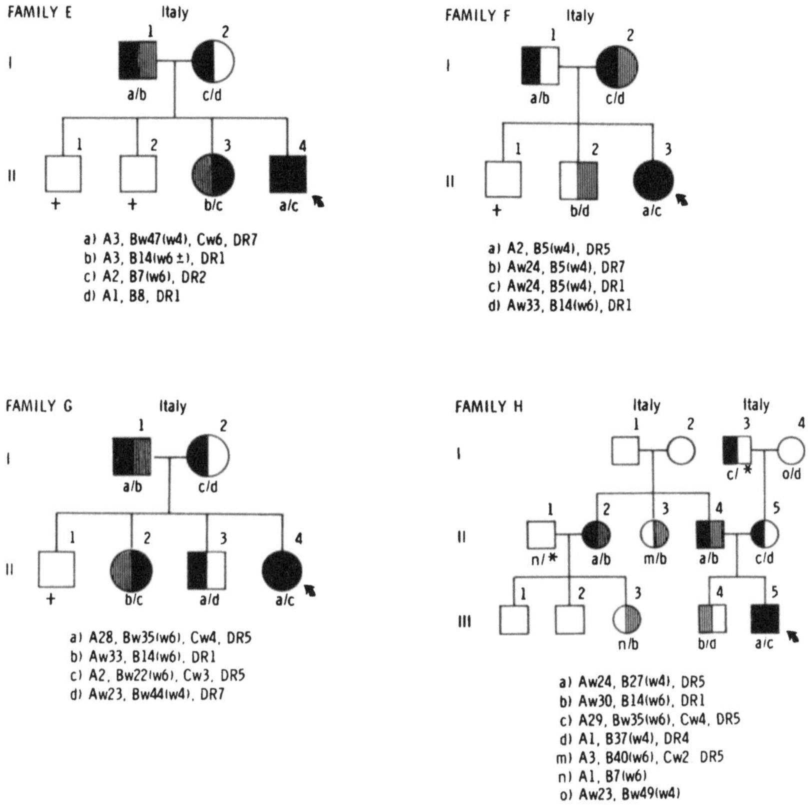

Fig. 15 Pedigrees of families E-H with classical and asymptomatic non-classical 21-hydroxylase deficiency. See Fig. 14 for legend.

ciency, despite the fact that both patient groups demonstrated similar biochemical abnormalities. In the group with symptomatic non-classical 21-hydroxylase deficiency, the excessive androgen levels produced virilization, while in the other group, no such hormonal effect was observed. Patients with the symptomatic non-classical disorder were detected because of their clinical symptoms, whereas asymptomatic non-classical 21-hydroxylase deficiency was detected only as a result of genetic and hormonal studies of classical CAH families.

5. Hormonal standards for genotyping 21-hydroxylase deficiency. We have developed a series of nomograms relating the baseline and ACTH-stimulated levels of 17-OH-P, Δ^4, DHEA, the ratio DHEA/Δ^4, and testosterone (T). These nomograms provide hormonal standards to use in the assignment of the 21-hydroxylase deficiency genotype; i.e. patients whose hormonal values fall on the regression line within a defined group are assigned to that group (Figs. 16-21). Clinical symptoms and signs must distinguish between the asymptomatic and symptomatic non-classical forms of 21-hydroxylase deficiency, which are biochemically indistinguishable. Heterozygotes for classical, symptomatic non-classical and asymptomatic non-classical 21-hydroxylase deficiencies demonstrate a similar hormonal response and are thus phenotypically indistinguishable. Both the asymptomatic and the symptomatic non-classical forms of 21-hydroxylase deficiency can occur in individuals with two mild deficiency alleles, or with one mild deficiency allele and one severe, classical deficiency allele. The homozygous and heterozygous non-classical 21-hydroxylase deficiency genotypes are hormonally indistinguishable and must be determined by family studies. The distribution of hormonal responses along a regression line suggests a spectrum of enzymatic deficiency in these groups: patients with classical CAH have the most severe deficiency; patients with non-classical forms have a less severe deficiency; while heterozygotes for all three forms have an even milder deficiency unmasked only upon ACTH stimulation. Hormonal values in family members predicted by HLA genotyping to be unaffected for 21-hydroxylase deficiency fall at the lowest point of the regression line, and serve as the best control population for normal 21-hydroxylase activity. Those members of the general population whose responses are in the heterozygote range may actually be carriers of a gene for 21-hydroxylase deficiency.

6. Allelic variants. In 1973, McKusick (125) listed CAH due to 21-hydroxylase deficiency among the disorders in which the phenotypic diversity might be attributed to allelic series.

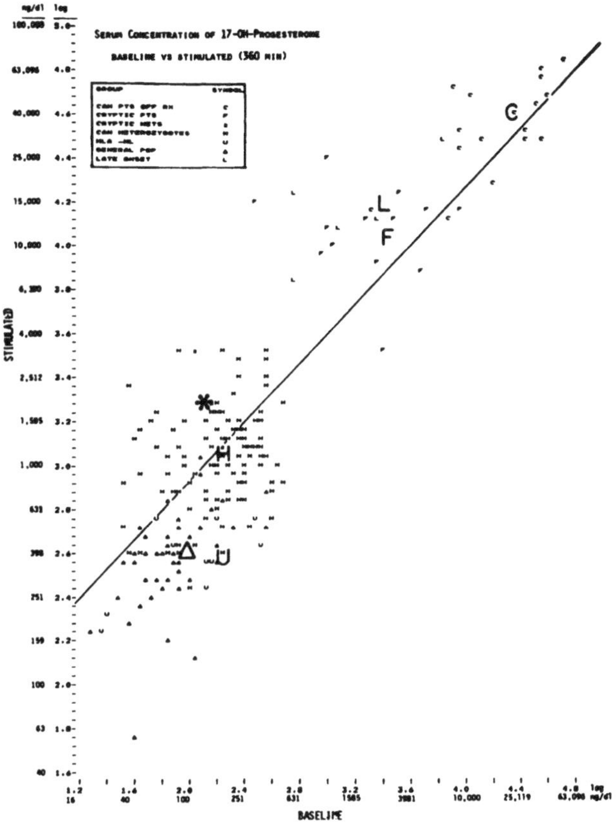

Fig. 16 Nomogram relating baseline to the ACTH stimulated serum 17-hydroxyprogesterone concentration. Values for $\log_{10}$ and antilog are indicated. The mean for each group is indicated by a large symbol.

C = Classical CAH patients off treatment
F = Asymptomatic non-classical 21-hydroxylase deficiency patients
* = Heterozygotes for non-classical 21-hydroxylase deficiency
H = Heterozygotes for classical CAH
U = Family members predicted by HLA genotyping to be genetically unaffected
# = General population
L = Symptomatic non-classical 21-hydroxylase deficiency patients

The values for each group aggregate in descending

order along the regression line. Note that the heterozygotes for the non-classical 21-hydroxylase defects are in the same range as the heterozygotes for the classical 21-hydroxylase defect. (Figs. 16-21 from New et al. (56), by permission of Academic Press)

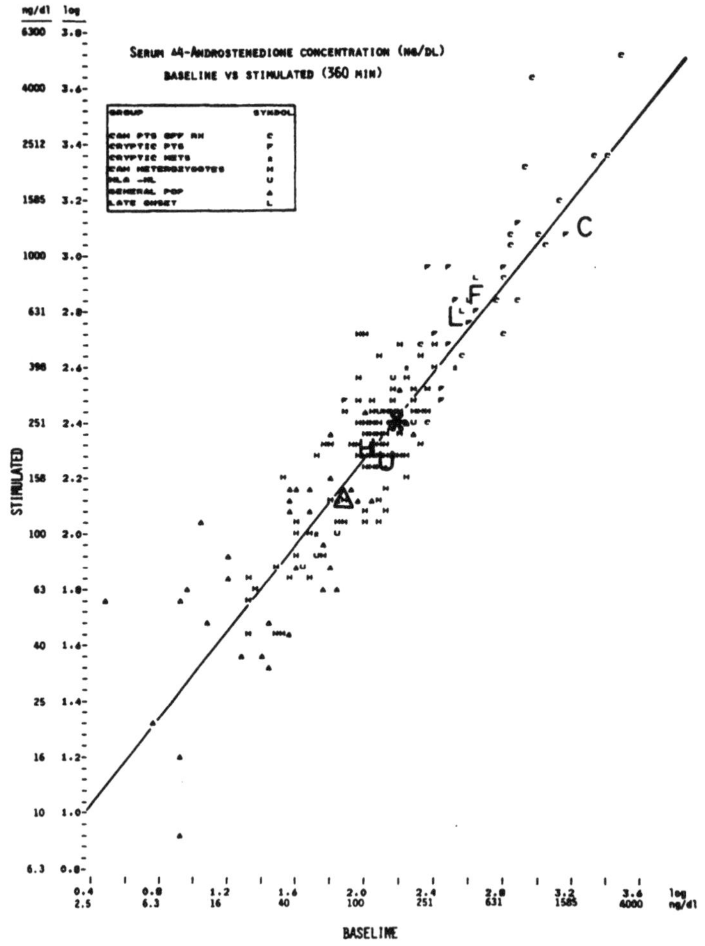

Fig. 17 Nomogram relating baseline to ACTH stimulated serum Δ^4 -androstenedione concentration. Values for $\log_{10}$ and antilog are indicated. The mean for each group is indicated by a large symbol. See legend to Fig. 16 for symbols. The values for each group aggregate

in descending order along the regression line. Note that the values for the asymptomatic patients with non-classical 21-hydroxylase deficiency are in the same range as the symptomatic virilized patients with non-classical 21-hydroxylase deficiency.

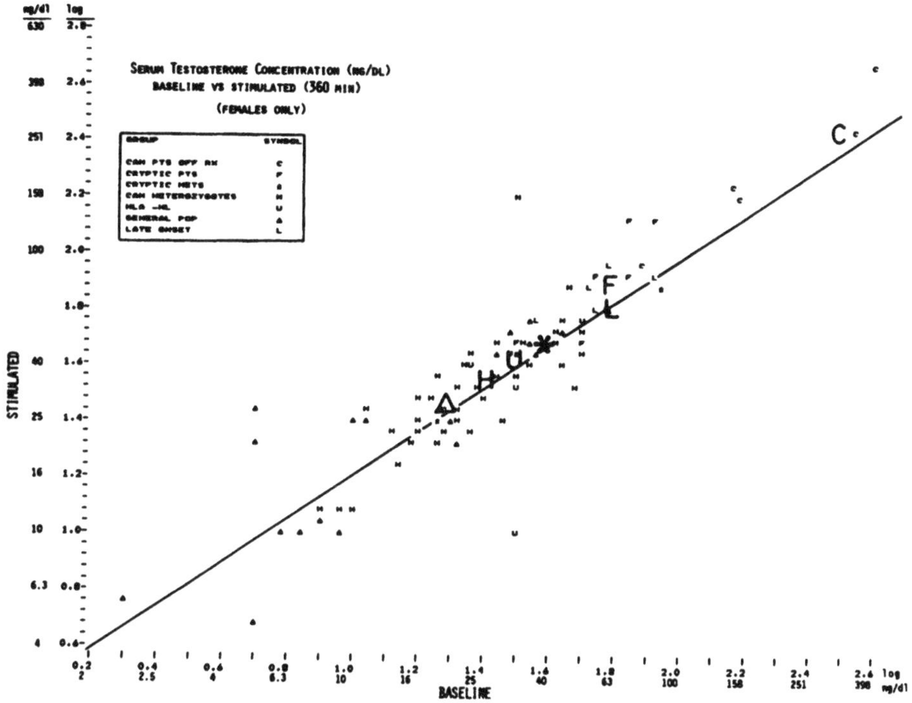

Fig. 18 Nomogram relating baseline to ACTH stimulated serum testosterone concentration in females. Values for $\log_{10}$ and antilog are indicated. The mean for each group is indicated by a large symbol. See legend to Fig. 16 for symbols. The values for each group aggregate in descending order along the regression line. Note that the values for the asymptomatic females with non-classical 21-hydroxylase deficiency are in the same range as the symptomatic virilized females with non-classical 21-hydroxylase deficiency.

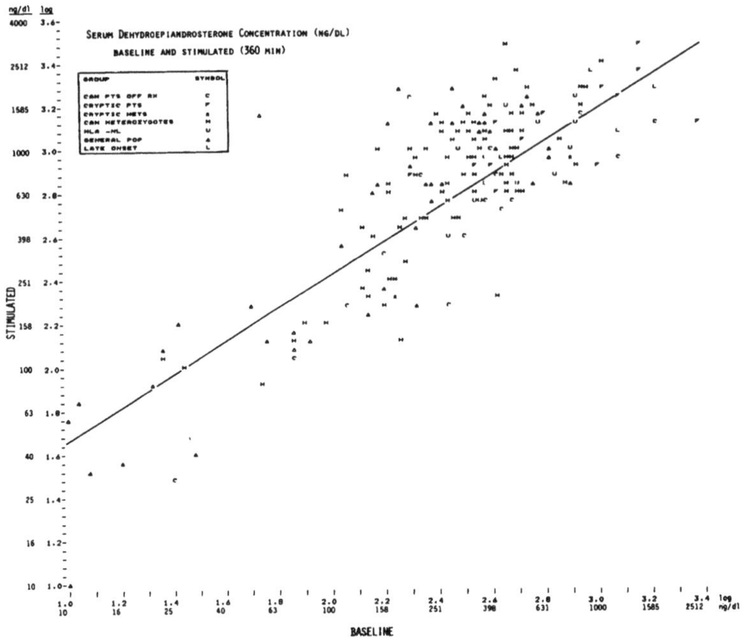

Fig. 19 Nomogram relating baseline to ACTH stimulated serum dehydroepiandrosterone concentration. Values for $\log_{10}$ and antilog are indicated. See legend to Fig. 16 for symbols. Note that the serum dehydroepiandrosterone concentration does not aggregate according to group.

McKusick noted that genetic compounds were an additional source of phenotypic diversity. Although the phenotypic variants of salt-wasting and simple virilizing CAH due to 21-hydroxylase deficiency have long been recognized, the new phenotypic variant of asymptomatic non-classical 21-hydroxylase deficiency has only recently been described (122).

We propose that there are allelic variants at the 21-hydroxylase locus which produce different degrees of 21-hydroxylase deficiency resulting in the phenotypic diversity of

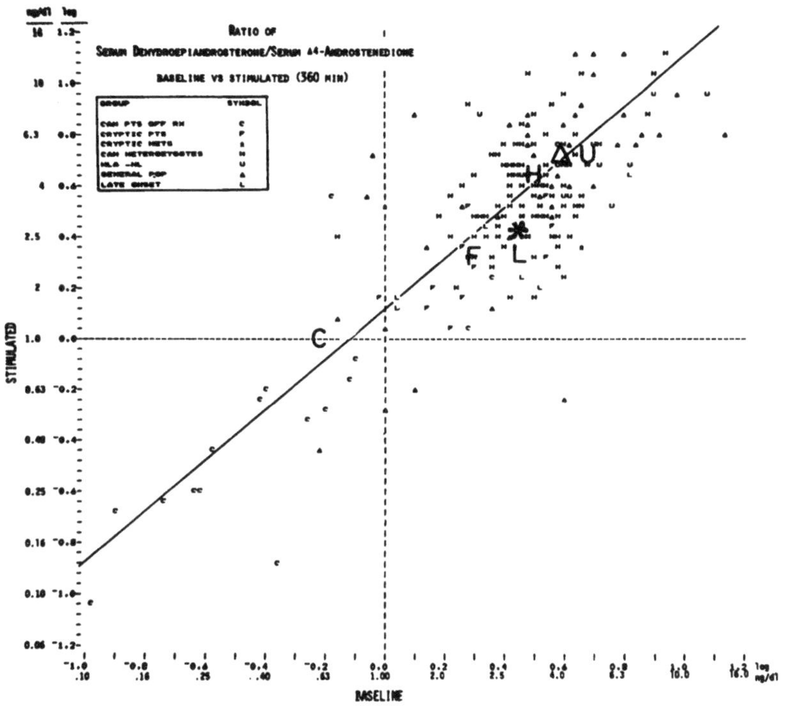

Fig. 20 Nomogram relating baseline to ACTH stimulated ratio of serum dehydroepiandrosterone/Δ^4-androstenedione concentrations. Values for $\log_{10}$ are indicated. The mean for each group is indicated by a large symbol. See legend to Fig. 16 for symbols. The values for each group aggregate in descending order along the regression line. Note that the ratio of DHEA/Δ^4 is lower in patients with classical congenital adrenal hyperplasia, and that this group is easily distinguished from other groups by the ratio.

classical, symptomatic non-classical and asymptomatic non-classical 21-hydroxylase deficiencies. We further suggest that the non-classical form may be due to either a homozygous non-classical deficiency genotype, or a genetic compound of the classical (or severe) 21-hydroxylase deficiency allele and the mild non-classical 21-hydroxylase deficiency allele.

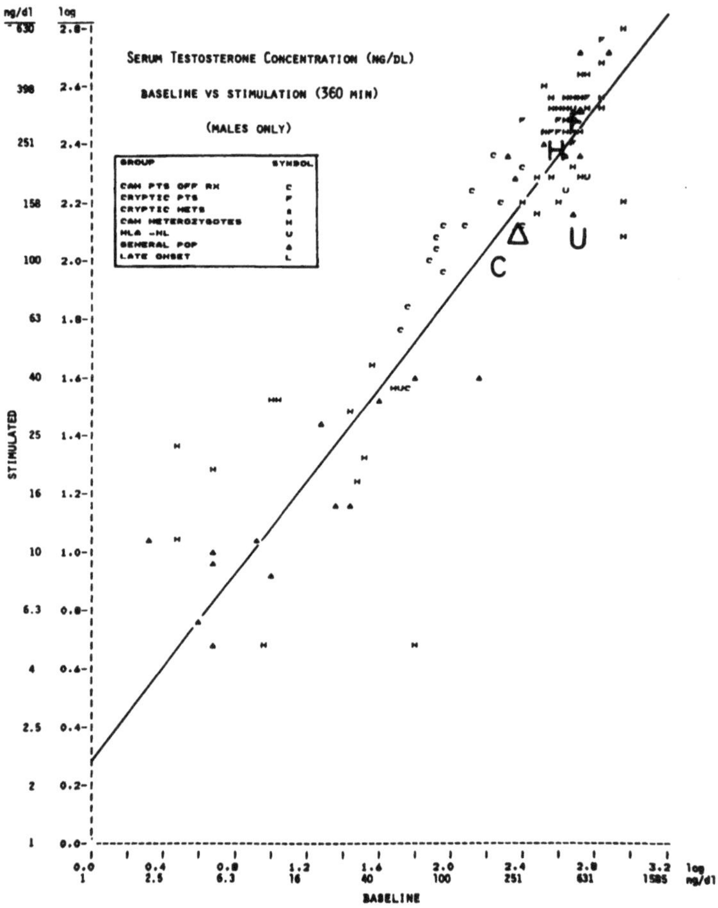

Fig. 21 Nomogram relating baseline to the ACTH stimulated serum testosterone concentration in males. Values for $\log_{10}$ and antilog are indicated. The mean for each group is indicated by a large symbol. See legend to Fig. 16 for symbols. Note that in males the serum testosterone does not distinguish the groups as well as in females.

This postulate awaits proof by complementation studies. Ultimate proof may be obtained by sequencing the gene by recombinant DNA techniques and demonstrating the specific nature of the genetic mutation for each allelic variant.

One may anticipate that a similar degree of genetic heterogeneity is likely to be found in the other inherited enzyme defects of steroidogenesis. Thus, the phenotypic variability already recognized in the 11β-hydroxylase, 3β-HSD and 17β-HSD deficiency may be examples of genetic heterogeneity.

7. Genetic linkage disequilibrium. An important distinction in genetic studies is the distinction between genetic linkage, which is described above for HLA and 21-hydroxylase deficiency, and genetic linkage disequilibrium, which is the non-random association of particular alleles of different genetic loci. In patients with classical and non-classical 21-hydroxylase deficiencies, certain HLA antigens appear with either a significantly increased or decreased frequency, relative to their frequency in the general population. Thus, not only are 21-hydroxylase deficiency and HLA genetically linked, but there is genetic linkage disequilibrium between the 21-hydroxylase deficiency and specific HLA genes.

The most significant association for classical 21-hydroxylase deficiency has been found for HLA-Bw47, where the combined relative risk for Caucasians from different geographical areas is 15.4 (105). Slight increases have also been reported for Bw51, Bw53, Bw60 and DR7. A review of the Bw47 positive haplotypes in patients with 21-hydroxylase deficiency reveals that this antigen frequently occurs on one particular haplotype: A3, Cw6, Bw47, DR7. Although the gene frequency for Bw47 in different Caucasian populations is always very low (<0.005), Bw47 appears with remarkable frequency among 21-hydroxylase CAH patients; in one particular region of England, the Bw47 frequency among CAH patients is nearly 50 per cent. Several studies have also demonstrated that A1, B8 and DR3 are consistently decreased among 21-hydroxylase patients (104, 110, 126-128).

Genetic linkage disequilibrium has also been reported for

the non-classical forms of 21-hydroxylase deficiency (107, 118, 119). The latest study (121) documents the significantly increased frequency of HLA-B14, DR1 and complement factor BfS in non-classical 21-hydroxylase deficiency. These alleles tend to appear together on the same haplotype in patients with non-classical 21-hydroxylase deficiency, suggesting that the HLA haplotype segment HLA-B14, DR1, BfS, rather than the individual HLA alleles, is highly associated with this form of 21-hydroxylase deficiency.

The analysis of genetic linkage disequilibrium provides provocative evidence that some patients with symptomatic non-classical 21-hydroxylase deficiency, similar to patients with asymptomatic non-classical 21-hydroxylase deficiency, may be genetic compounds possessing both a severe and a mild 21-hydroxylase deficiency gene. A few symptomatic non-classical patients have the HLA type HLA-Bw47, an allele occurring with strikingly high frequency in association with the gene for classical 21-hydroxylase deficiency. In the family depicted in Fig. 22, for example, the appearance of Bw47 on haplotype "c" makes the presence of a classical deficiency gene highly likely. Because heterozygotes for the classical gene and heterozygotes for the non-classical gene are biochemically indistinguishable, hormonal tests provide no indication of whether the mother (I-2) carries a mild or severe deficiency gene on haplotype "c". Conclusive proof for the presence of a classical 21-hydroxylase deficiency gene rests on the eventual appearance of a patient with classical 21-hydroxylase deficiency in the family pedigree. In studying the family of a patient recently diagnosed as having congenital salt-wasting 21-hydroxylase deficiency (Fig. 23), the maternal grandmother (I-4) had a hormonal response and medical history consistent with symptomatic non-classical 21-hydroxylase deficiency. Particularly noteworthy is the fact that the HLA haplotype which the grandmother shares with the patient with classical CAH (Fig. 23, haplotype "a") contains Bw47, while the grandmother's other HLA haplotype contains B14, the HLA antigen most highly associated with non-classical 21-

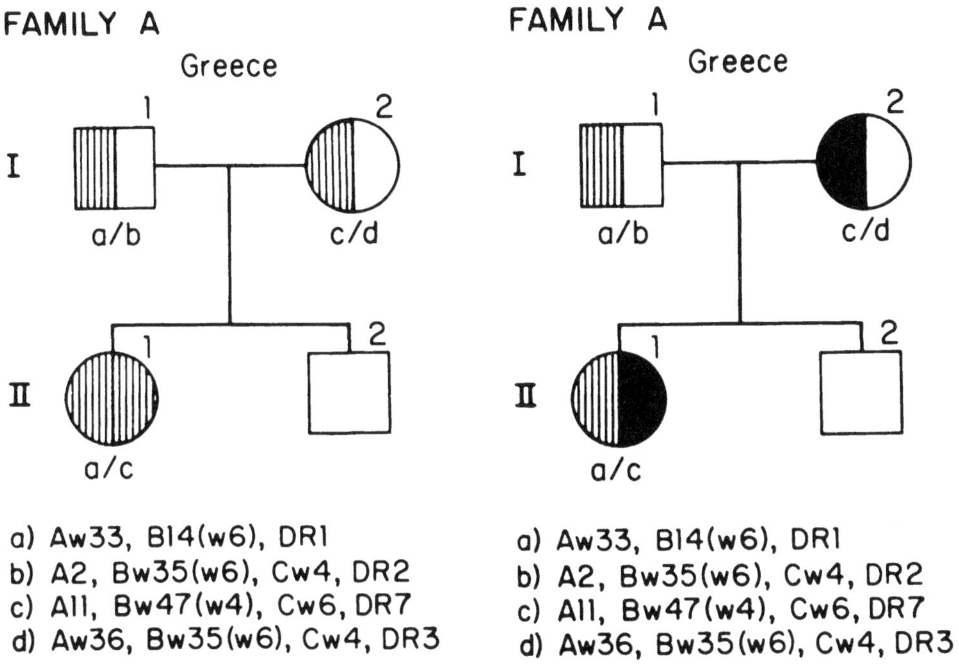

Fig. 22 Two possible genotypes for symptomatic non-classical 21-hydroxylase deficiency. On the left: Both of the patient's HLA haplotypes are linked to mild 21-hydroxylase deficiency alleles (striped symbols). On the right: The patient is a genetic compound, possessing a Bw47-linked classical 21-hydroxylase deficiency allele (dark symbol) in addition to a mild deficiency allele. Note that hormonal testing cannot distinguish between the two different heterozygote genotypes proposed for the mother. (Adapted from New et al. (153), by permission of McGraw-Hill).

hydroxylase deficiency.

The finding that both the symptomatic and asymptomatic non-classical 21-hydroxylase deficiency genes are associated with the same HLA haplotype segment (B14, DR1, BfS) suggests that these two variant genes are highly related. It is possi-

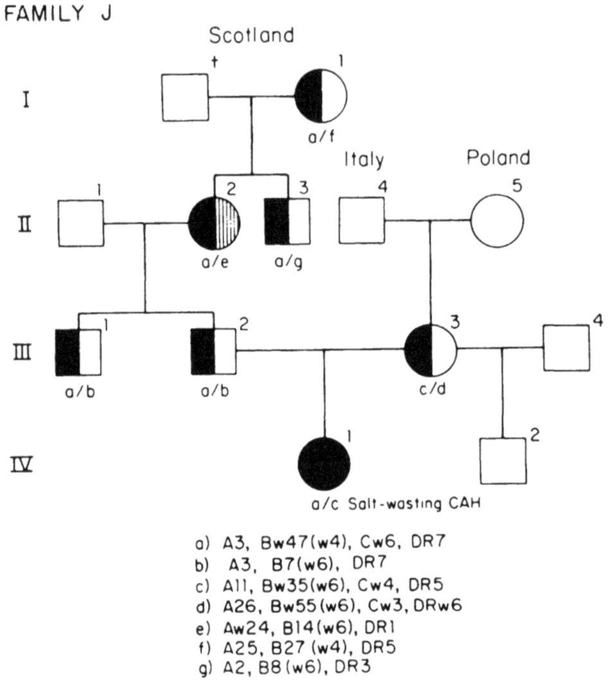

Fig. 23 Occurrence of classical and symptomatic non-classical 21-hydroxylase deficiency in the same family. It is proposed that the grandmother (II-2) with symptomatic non-classical 21-hydroxylase deficiency is a genetic compound, possessing both an HLA-Bw47-linked severe 21-hydroxylase deficiency allele (solid symbol) and an HLA-B14-linked mild deficiency allele (striped symbol). (Adapted from New et al. (153), by permission of McGraw-Hill).

ble that individuals who have highly similar or identical non-classical 21-hydroxylase deficiency alleles may express different clinical manifestations of their 21-hydroxylase defect, expressing either the symptomatic or asymptomatic disorder. Similarly, the clinical distinction between the symptomatic and asymptomatic disorders no longer appears as discrete as initially appreciated. Marked variability was found in the symptoms of the non-classical patients recently examined, ranging from precocious pubic hair growth without subsequent symptoms of virilization, to more severe degrees of hirsutism, acne, menstrual disturbances and stunted growth.

8. Clinical spectrum of 21-hydroxylase deficiency - phenotypic variability. It thus appears that 21-hydroxylase deficiency has a spectrum of clinical manifestations (Fig. 24) as well as a spectrum of enzymatic deficiency, as reflected in the hormonal findings presented in Figs. 16-21. If 21-hydroxylase deficiency is similar to other autosomal recessive enzyme defects, further studies should reveal patients whose clinical and biochemical abnormalities will be intermediate between those currently described. It is also possible that the assignments made by clinical criteria may change in the course of a lifetime. In one patient (Fig. 25), the symptoms of mild 21-hydroxylase deficiency, including pubic hair and advanced growth, were evident at six months of age. The symptoms subsequently disappeared but the biochemical abnormality persisted. Thus, the clinical classification for this patient changed from symptomatic to asymptomatic 21-hydroxylase deficiency within a span of ten years.

B. The genetics of other enzyme defects of steroidogenesis

The 11β-hydroxylase, 3β-HSD and 17α-hydroxylase deficiencies are also transmitted as autosomal recessive traits, but are rarer than the 21-hydroxylase deficiency. The mode of inheritance of the other enzyme defects has not been documented.

Studies of families of patients with CAH due to 11β-hydroxylase deficiency (105) and 17α-hydroxylase deficiency (129) have demonstrated that these enzyme defects are not genetically linked to HLA. In addition, hormonal studies have not provided a means of detecting heterozygote carriers for 11β-hydroxylase deficiency (130). One study of corticosterone methyl oxidase Type II deficiency, failed to demonstrate close

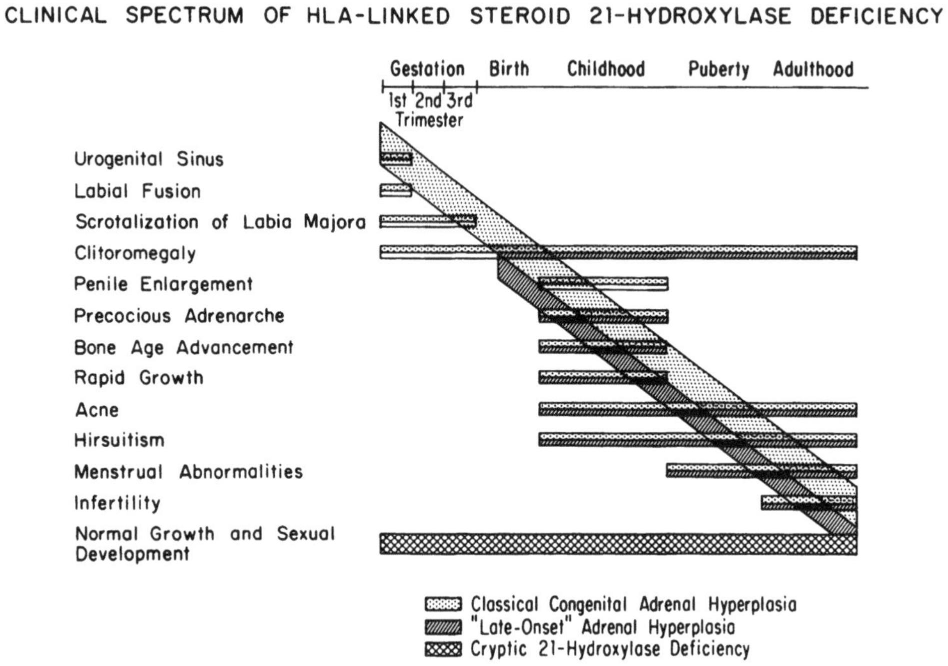

Fig. 24 Clinical spectrum of 21-hydroxylase deficiency. There is a wide spectrum of clinical presentation in 21-hydroxylase deficiency ranging from prenatal virilization with labial fusion to precocious adrenarche to pubertal or post-pubertal virilization. During the course of a lifetime a patient may change from a symptomatic to an asymptomatic patient with 21-hydroxylase deficiency. (From New et al. (153), by permission of McGraw-Hill).

genetic linkage with HLA, although a LOD Score of 1.128 was reported, and thus the possibility of genetic linkage could not be ruled out (131).

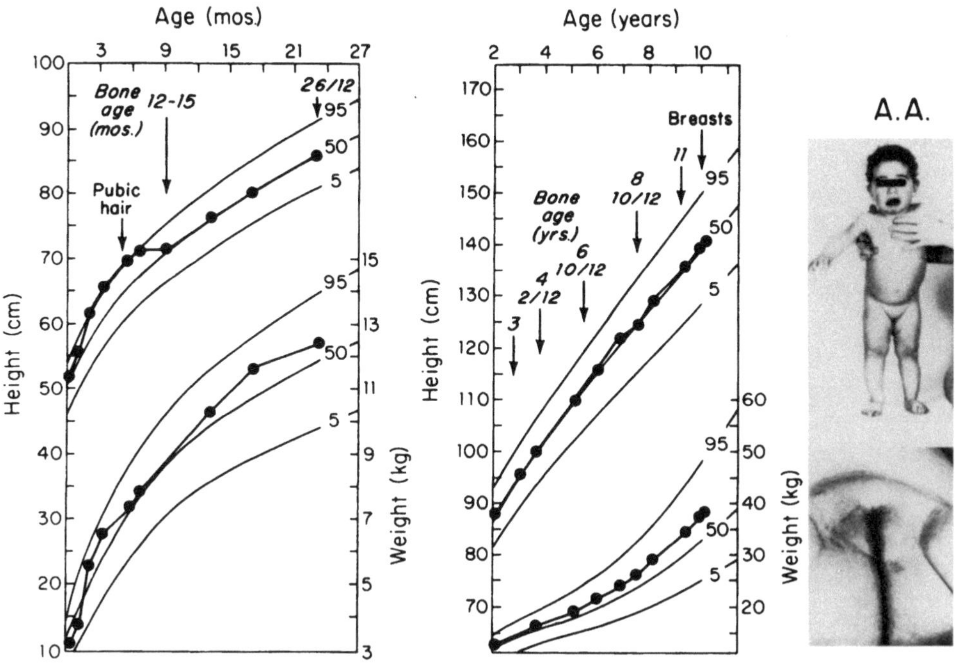

Fig. 25 The changing clinical profile from symptomatic to asymptomatic non-classical 21-hydroxylase deficiency. In this female, postnatal onset of virilization and growth acceleration was followed by a remission of symptoms during childhood, despite persistance of hormonal abnormalities. (From New et al. (153), by permission of McGraw-Hill).

C. Neonatal screening for 21-hydroxylase deficiency CAH

In 1977, neonatal screening for 21-hydroxylase CAH became possible by the development of a microfilter paper method for measuring 17-OH-P (132). This method utilizes a heel stick blood specimen which is spotted onto filter paper and then analyzed for 17-OH-P by radioimmunoassay. Results are shown in Fig. 26. The hormone is stable on filter paper and the filter

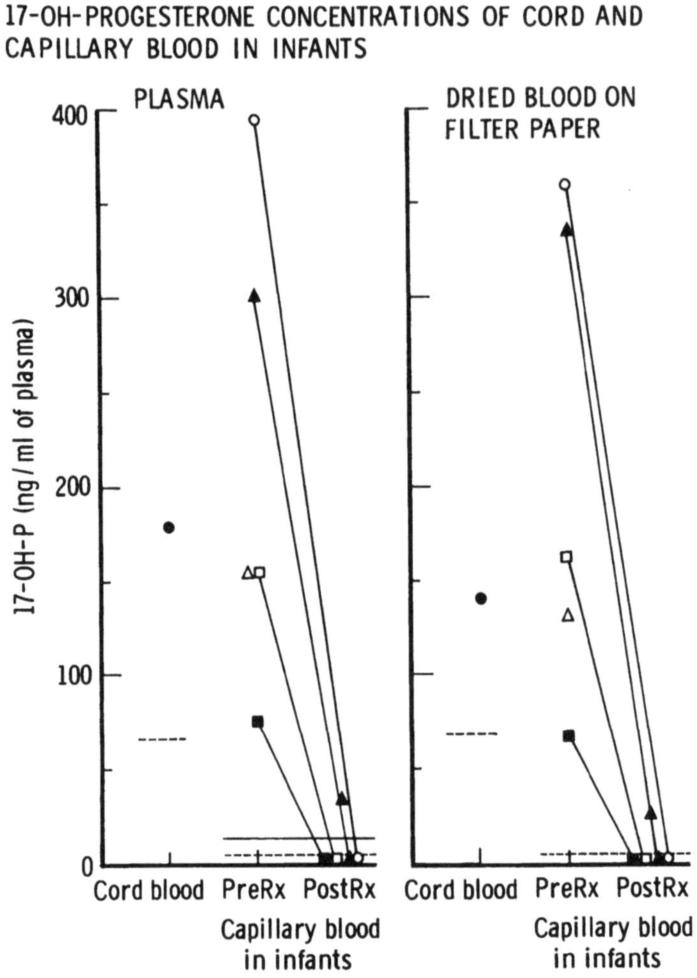

Fig. 26 17-OH Progesterone concentration in cord and capillary blood samples in infants with CAH. Ages of infants at diagnosis were as follows:

● = cord blood; Δ = 2 days; □ = 4 days; ■ = 7 days; ▲ = 2 weeks; O = 4 weeks. Samples on treatment were taken from 5 days to 1 month after treatment was begun. The dashed line indicates upper limit of normal infants. The solid line indicates upper limit of pooled plasma in sick infants. Plasma samples were pipetted quantitatively and assayed. Cord blood samples were also quantitatively pipetted onto filter paper. Capillary samples were directly applied to filter paper without quantitation and a 3 mm. disc was punched out for analysis. (From Pang et al. (132), by permission of The Endocrine Society).

paper specimens can be sent to an appropriate laboratory by surface mail. Based on population surveys which indicate that the incidence of 21-hydroxylase deficiency is approximately 1:15,000 before screening (an incidence equal to that of phenylketonuria after screening), it was proposed that the newborn population be screened for 21-hydroxylase deficiency (133). A number of investigators have reported individuals with classical 21-hydroxylase deficiency who have gone unrecognized (109, 134).

The reliability and feasibility of the 17-OH-P microfilter paper screening method has recently been proven by a pilot study screening all infants born in Alaska in a 30-month period (135). A total of 19,677 consecutive newborns were screened on the third day of life. Four of 1,131 newborn Yupik Eskimos screened and 1 of 13,733 newborn Caucasians screened were found to be affected with CAH (Table 3). All 5 affected patients had the salt-wasting form of CAH. Based on these data, a 95 percent confidence limit establishes a range for the incidence of salt-wasting CAH of between 1:110-1:1,038 live births in Yupiks and 1:2,465-1:524,000 live births in Caucasians (Table 3). The incidence of salt-wasting CAH in Yupiks agrees with the incidence previously predicted by case survey (1:292-1:896) in this population (102). In Caucasians, the highest incidence of salt-wasting CAH allowed at the 95 percent confidence level based on screening data (1:2,465) is far greater than the highest possible incidence predicted by case surveys (1:18,454 for salt-wasting CAH and 1:14,798 for salt-wasting and simple virilizing CAH combined) (Table 3). A larger screening sample size is needed to permit more definite conclusions about the frequency of this disorder. The false positive and recall rates for the pilot screening program were 0.05% and 0.1% respectively (135),

Table 3. Frequency of CAH Gene Based on Newborn Screening and Case Surveys in Alaska

	Cases	Sample	Homozygous Affected Population: Estimated Incidence	Homozygous: 95th Percent Conf. Limit of Est Rate* Lower	Homozygous: Upper	Heterozygous Population: Estimated Incidence	Heterozygous: 95th Percent Conf. Limit of Est Rate* Lower	Heterozygous: Upper
YUPIK ESKIMO (salt-wasting only)								
Alaska screening (a)	4	1,131	1:282	1:110	1:1038	1:8.9	1:5.8	1:16.6
Case survey (b)	14	6,860	1:490	1:292	1:896	1:11.6	1:9.0	1:15.5
CAUCASIAN								
Alaska screening (salt-wasting only)	1	13,733	1:13,733	1:2465	1:524,000	1:59	1:25	1:362
Case survey (c-g) (salt-wasting only)	106	2,343,895	1:22,112	1:18,454	1:26,730	1:74.8	1:68	1:82
Case survey (c-e, g-k) (all 21-hydroxylase deficient)	215	3,619,891	1:16,836	1:14,798	1:19,239	1:65.4	1:61	1:69.8

* Confidence limit for the expectation of Poisson variable for the estimated rate.

a Pang et al., 1982 (135)
b Hirschfeld and Fleshman, 1969 (102)
c Muller et al., 1979 (99)
d Werder et al., 1980 (133)
e Rosenbloom and Smith, 1966 (98)
f Qazi and Thompson, 1972 (150)
g Mauthe et al., 1980 (151)
h Childs et al., 1956 (95)
i Prader, 1958 (96)
j Wilkins, 1962 (20)
k Hubble, 1966 (149)

figures that compare favorably with the rates in currently enacted screening programs for other disorders (136).

Two additional neonatal screening programs have been completed. In Rome, Natoli et al. (137) found 4 cases out of 22,400 births for an incidence of 1:5,580 cases; in Bologna, Cacciari et al. (138) found 5 cases out of 42,930 births for an incidence of 1:8,586 cases.

D. Prenatal diagnosis of CAH

1. <u>21-Hydroxylase deficiency</u>. Since the report by Jeffcoate et al. (139) of the prenatal diagnosis of CAH by elevated concentrations of 17-ketosteroids and pregnanetriol in the amniotic fluid of the affected fetus, several investigators have attempted prenatal diagnosis by measurement of various hormones (22). Most recently, elevated levels of 17-OH-P (140-143) and Δ^4(143) in the amniotic fluid of fetuses affected with CAH due to 21-hydroxylase deficiency have been reported (Fig. 27).

The discovery of genetic linkage between the HLA complex and the 21-hydroxylase deficiency gene (103) has provided an additional method for prenatal diagnosis of 21-hydroxylase deficiency in a pregnancy at risk. HLA genotyping of amniotic cells has made possible the prediction of a heterozygous, affected or unaffected fetus (144). When HLA genotyping of amniotic cells reveals that the fetus is HLA-identical to the affected sib, the fetus is predicted to be affected. The HLA prediction of CAH genotype should always be corroborated by hormonal measurement of 17-OH-P and Δ^4 in amniotic fluid. Caution must be exercised in interpreting results if multiple births are expected or if there is antigen sharing in the parents (145).

2. <u>11β-Hydroxylase deficiency</u>. Recently, levels of 11-deoxycortisol and tetrahydrocortisol (THS) in amniotic fluid and THS in maternal urine have been found to be increased in pregnancies with fetuses affected with 11β-hydroxylase deficiency (146, 147). This suggests that prenatal diagnosis of this disorder by hormonal measurement is feasible. Since this disorder

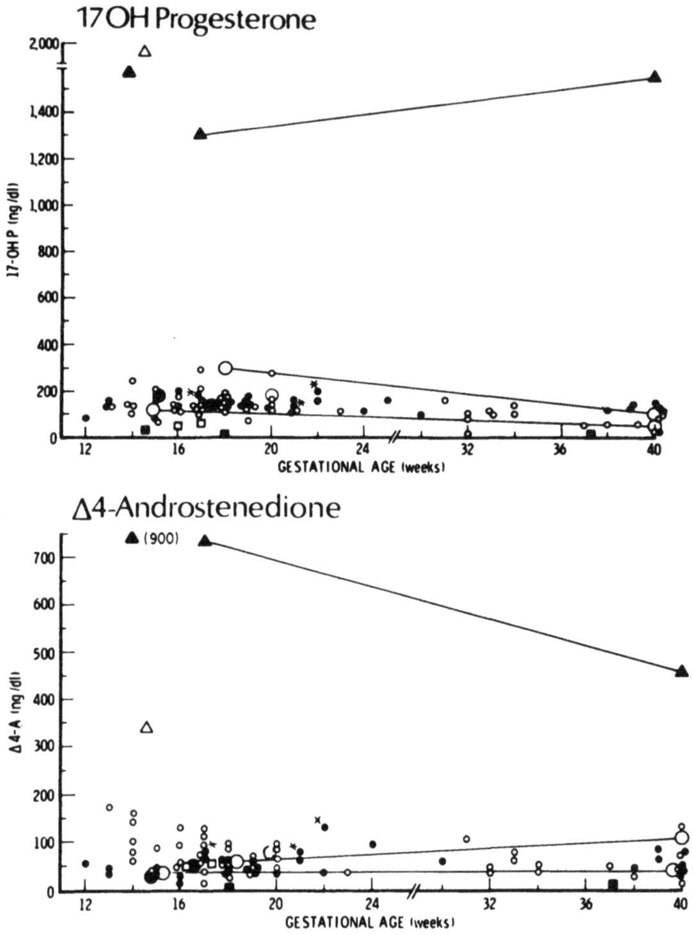

Fig. 27 17-OHP and Δ^4 concentrations in amniotic fluid samples of normal, CAH and anencephalic fetuses from 12 weeks of gestational age to term. The lines between two values indicate specimens from the same fetuses. *, Amniotic fluid contaminated with blood. (From Pang et al. (143), by permission of The Endocrine Society).

is not genetically linked to HLA, HLA typing of amniotic cells is not helpful.

E. Sex assignment

Sexual ambiguity at birth is a medical emergency. In such cases, a rational and judicious choice of sex assignment is a critical aspect of treatment, since the sex assignment has obvious life-long implications. Determination of genetic sex by karyotype or buccal smear and the accurate diagnosis of the specific underlying enzymatic defect are essential in assessing a patient's potential for future sexual activity and fertility.

In cases of female pseudohermaphroditism due to 21- or 11β-hydroxylase deficiency, a female sex assignment is appropriate. When medical treatment is begun early in life, the initially large and prominent clitoris shrinks slightly and, as the surrounding structures grow normally, it becomes much less prominent and may not require surgical revision. When the clitoris is conspicuously enlarged or when the abnormal genitalia interfere with parent-child bonding, plastic surgery to correct the appearance of the clitoris should be carried out. Definitive vaginoplasty should, in general, be performed in early childhood by an experienced gynecological surgeon (57, 148). Because of the normal internal genitalia and gonads in these patients, normal puberty, fertility and childbearing are possible when there is early therapeutic intervention.

In cases of male pseudohermaphroditism due to enzyme deficiencies impairing androgen synthesis, a sex assignment consistent with the genetic sex -- i.e. a male sex assignment -- is not always optimal. Virilization of the genitalia in these children is frequently so extremely and irrevocably incomplete that the anatomy precludes normal male functioning. There are certain physiological capacities we consider integral to "normal" male sexual development: a capacity for intercourse, for urinating in a standing position, and for developing secondary sex characteristics adequate for the presentation of a male gender in society. Because these capacities are often severely compromised in genetic males with congenital disorders of sex steroid synthesis, a female sex assignment may be advisable. If the male pseudohermaphrodite is to be raised as a female, surgical correction of the genitalia and gonadectomy are required; with appropriate therapeutic measures, relatively normal, albeit infertile, female sexual activity and development are possible. In cases of impaired androgen synthesis, administration of sex steroids is usually required to induce development of appropriate sexual characteristics at puberty, either estrogens if the patient is reared as a female or androgens if reared as a male.

Society sees phenotype, not genotype. In assigning a sex of rearing to a male or female pseudohermaphrodite, the genetic sex is of less consequence than the physiological and anatomical character of the genitalia and their potential for development and function. Because of the wide variation in the presentation of ambiguous genitalia in these patients, there can be no all-inclusive rules for sex assignment based solely on genetic sex or type of enzyme deficiency.

Summary

Several disorders of steroid biosynthesis are discussed and their hormonal and clinical manifestations, as well as current methods of diagnosis and treatment are described. The recent discovery of genetic linkage between CAH due to 21-hydroxylase deficiency and HLA has been a major development in the study of CAH. The use of HLA genotyping as a marker for the CAH genotype has been applied to better heterozygote detection, prenatal diagnosis of CAH, and more sophisticated characterization of the various forms of 21-hydroxylase deficiency. It is proposed that a spectrum of 21-hydroxylase deficiencies exist with differing degrees of severity of enzyme deficiency. Further hormonal and immunogenetic studies should prove helpful in elucidating the biological basis for the variable expression of these disorders of adrenal steroidogenesis.

Acknowledgements

Much of the work described in this review was supported by NIH grants: HD 00072, HD 15084, HL 17749, CA 22507, CA 08747, CA 19267, AI 15227, and by grant RR 47 from the General Clinical Research Centers Program of the Division of Research Resources, NIH.

References

1. DeCrecchio, L.: Sopra un caso di apparenze virile in una donna. Morgagni 7, 1951 (1865).
2. Bongiovanni, A.M., Eberlein, W.R., Goldman, A.S., New, M.I.: Disorders of adrenal steroid biogenesis. Recent Prog. Horm. Res. 23, 375-439 (1967).
3. Finkelstein, M., Shaefer, J.M.: Inborn errors of steroid biosynthesis. Physiol. Rev. 59, 353-406 (1979).
4. Gaunt, R.: Action of adrenal cortical steroids on electrolyte and water metabolism. In: "The Human Adrenal Cortex", Ed. Christy, N.P., Harper and Row, New York, pp. 273-302 (1971).
5. Ganong, W.F.: The central nervous system and the synthesis and release of adrenocorticotrophic hormone. In: "Advances in Neuroendocrinology", Ed. Nalbandov, A.B., University of Illinois Press, Urbana, pp. 92-149 (1963).
6. Guillemin, R., Schally, A.V.: Recent advances in the chemistry of neuroendocrine mediators originating in the central nervous system. In: "Advances in Neuroendocrinology", Ed. Nalbandov, A.B., University of Illinois Press, Urbana, pp. 314-328 (1963).
7. Laragh, J.H.: Aldosteronism in man: Factors controlling secretion of the hormone. In: "The Human Adrenal Cortex", Ed. Christy, N.P., Harper and Row, New York, pp. 483-507 (1971).
8. Mueller, J.: "Regulation of Aldosterone Biosynthesis", Springer-Verlag, New York (1971).
9. New, M.I., Seaman, M.P.: Secretion rates of cortisol and aldosterone precursors in various forms of congenital adrenal hyperplasia. J. clin. Endocrinol. Metab. 30, 361-371(1970).

10. Ganong, W.F., Biglieri, E.G., Mulrow, P.J.: Mechanisms regulating adrenocortical secretion of aldosterone and glucocorticoids. Recent Prog. Horm. Res. 22, 381-414 (1966).

11. Ganong, W.F., Alpert, L.C., Lee, T.C.: ACTH and the regulation of adrenocortical secretion. New Engl. J. Med. 290, 1006-1011 (1974).

12. Mason, P.A., Fraser, R., Semple, P.F., Morton, J.J.: The interaction of ACTH and angiotensin II in the control of corticosteroid plasma concentration in man. J. Steroid Biochem. 10, 235-239 (1979).

13. Jost, A.: Steroids and sex differentiation of the mammalian foetus. Excerpta med. Int. Cong. Ser. 132, 74-81 (1966).

14. Josso, N.: Permeability of membranes to the Mullerian-inhibiting substance synthesized by the human fetal testis *in vitro*. J. clin. Endocrinol. Metab. 34, 265-270 (1972).

15. Federman, D.D.: Disorders of fetal endocrinology: female pseudohermaphroditism. In: "Abnormal Sexual Development", Ed. Laron, Z., W.B. Saunders, Philadelphia, pp. 121-133 (1968).

16. Virdis, R., Saenger, P., Senior, B., New, M.I.: Endocrine studies in a pubertal male pseudohermaphrodite with 17-ketosteroid reductase deficiency. Acta Endocrinol. 87, 212-224 (1978).

17. Syndor, K.L., Kelley, V.C., Raile, R.B., Ely, R.S., Sayers, G.: Blood adrenocorticotrophin in children with congenital adrenal hyperplasia. Proc. Soc. exp. Biol Med. 82, 695-697 (1953).

18. Binoux, M., Pham-Huu-Trung, M.T., Gourmelen, M., Girard, F., Canlrobe, P.: Plasma ACTH in adrenogenital syndrome. Acta Pediatr. Scand. 61, 269-270 (1972).

19. Butler, G.C., Marrian, G.F.: Isolation of pregnane-3,17,20-triol from urine of women showing adrenogenital syndrome. J. biol. Chem. 119, 565-572 (1937).

20. Wilkins, L.: Adrenal disorders. II. Congenital virilizing adrenal hyperplasia. Arch. Dis. Childh. 37, 231-241 (1962).

21. Berson, S.A., Yalow, R.S.: Radioimmunoassay of ACTH in plasma. J. clin. Invest. 47, 2725-2751 (1968).

22. New, M.I., Levine, L.S.: Congenital adrenal hyperplasia. In: "Advances in Human Genetics", Eds. Harris, H., Hirschhorn, K.K., Plenum Press, London, pp. 251-326 (1973).

23. Levine, L.S., Rauh, W., Gottesdiener, K., Chow, D., Gunczler, P., Rapaport, R., Schneider, B., New, M.I.: New studies of the 11β-hydroxylase and 18-hydroxylase enzymes in the hypertensive form of congenital adrenal hyperplasia. J. clin. Endocrinol. Metab. 50, 258-263 (1980).

24. Eberlein, W.R., Bongiovanni, A.M.: Plasma and urinary corticosteroids in the hypertensive form of congenital adrenal hyperplasia. J. biol. Chem. 223, 85-94 (1956).

25. Brown, R.D., Strott, C.A.: Plasma deoxycorticosterone in man. J. clin. Endocrinol. Metab. 32, 744-750 (1971).

26. Bongiovanni, A.M.: The adrenogenital syndrome with deficiency of 3β-hydroxysteroid dehydrogenase. J. clin. Invest. 41, 2086-2092 (1962).

27. Parks, G.A., Bermudez, J.A., Anast, C.S., Bongiovanni, A.M., New, M.I.: Pubertal boy with 3β-hydroxysteroid dehydrogenase defect. J. clin. Endocrinol. Metab. 33, 269-278 (1971).

28. Schneider, G., Genel, M., Bongiovanni, A.M., Goldman, A.S., Rosenfeld, R.L.: Persistent testicular Δ^5-isomerase-3β-hydroxysteroid dehydrogenase deficiency in the Δ^5-3β-HSD form of congenital adrenal hyperplasia. J. clin. Invest. 55, 681-690 (1975).

29. New, M.I.: Male pseudohermaphroditism due to 17α-hydroxylase deficiency. J. clin. Invest. 49, 1930-1941 (1970).

30. Madan, K., Schoemaker, J.: XY Females with enzyme deficiencies of steroid metabolism: A brief review. Hum. Genet. 53, 291-295 (1980).

31. Biglieri, E.G., Herron, M.A., Brust, N.: 17α-Hydroxylation deficiency in man. J. clin. Invest. 45, 1946-1954 (1966).

32. Mantero, F., Busnardo, B., Riondel, A., Veyrat, R., Aistoni, M.: Hypertension arterielle, alcalose hypokalemique et pseudohermaphroditisme male par deficit en 17α-hydroxylase. Schweiz. med. Wschr. 101, 38-43 (1971).

33. Prader, A., Gurtner, H.P.: Das Syndrom des Pseudohermaphroditismus masculinus bei Kongenitaler Nebennierenrinden Hyperplasia ohne Androgen-Uberproduktion. Helv. paediat. Acta 10, 397-412 (1955).

34. Camacho, A.M., Kowarski, A., Migeon, C.J., Brough, A.J.: Congenital adrenal hyperplasia due to a deficiency of one of the enzymes involved in the biosynthesis of pregnenolone. J. clin. Endocrinol. Metab. 28, 153-161 (1968).

35. Kirkland, R.T., Kirkland, J.L., Johnson, C., Horning, M., Librick, L., Clayton, G.W.: Congenital lipoid adrenal hyperplasia in an eight-year-old phenotypic female. J. clin. Endocrinol. Metab. 36, 488-496 (1973).

36. Hochberg, R.B., McDonald, P.D., Feldman, M., Lieberman, S.: Studies on the biosynthetic conversion of cholesterol into pregnenolone. J. biol. Chem. 249, 1277-1285 (1974).

37. Degenhart, H.J.: A study of the cholesterol splitting enzyme system in normal adrenals and in adrenal lipoid hyperplasia. Acta paediat. Scand. 60, 611-612 (1971).

38. Visser, H.K.A., Cost, W.S.: A new hereditary defect in the biosynthesis of aldosterone: urinary C21-corticosteroid pattern in three related patients with a salt-losing syndrome suggesting an 18-oxidation defect. Acta Endocrinol. 47, 589-612 (1964).

39. Ulick, S., Gautier, E., Vetter, K.K., Markello, J.R., Yaffe, S., Lowe, C.U.: An aldosterone biosynthetic defect in a salt-losing disorder. J. clin. Endocrinol. Metab. 24, 669-672 (1964).

40. Ulick, S.: Diagnosis and nomenclature of the disorders of the terminal portion of the aldosterone biosynthetic pathway. J. clin. Endocrinol. Metab. 43, 92-96 (1976).

41. Veldhuis, J.D., Kulin, H.E., Santen, R.J., Wilson, T.E., Melby, J.C.: Inborn error in the terminal step of aldosterone biosynthesis. New Engl. J. Med. 303, 117-121 (1980).

42. Rosler, A., Rabinowitz, D., Theodor, R., Ramirez, L., Ulick, S.: The nature of the defect in a salt-wasting disorder in Jews of Iran. J. clin. Endocrinol. Metab. 44, 279-291 (1977).

43. Saez, J.M., de Peretti, E., Morera, A.M., David, M., Bertrand, J.: Familial male pseudohermaphroditism with gynecomastia due to a testicular 17-ketosteroid reductase defect. J. clin. Endocrinol. Metab. 32, 604-610 (1971).

44. Goebelsmann, U., Horton, R., Mestman, J.H.: Arce, J.J., Nagata, Y., Nakamura, R.M., Thorneycroft, I.H., Mishell, D.R.: Male pseudohermaphroditism due to testicular 17β-hydroxysteroid dehydrogenase deficiency. J. clin. Endocrinol. Metab. 36, 867-879 (1973).

45. Givens, J.R., Wiser, W.L., Summit, R.L., Kerber, I.J., Andersen, R.N., Pittaway, D.E., Fish, S.A.: Familial male pseudohermaphroditism without gynecomastia due to deficient testicular 17-ketosteroid reductase activity. New Engl. J. Med. 291, 938-944 (1974).

46. Knorr, D., Bidlingmaier, F., Engelhardt, D.: Reifenstein's syndrome. Acta Endocrinol. Suppl. 173: 37 (1973).

47. Levine, L.S., Lieber, E., Pang, S., New, M.I.: Male pseudohermaphroditism due to 17-ketosteroid reductase deficiency diagnosed in the newborn period. Pediatr. Res. 14, 480 (1980).

48. Zachmann, M., Vollmin, J.A., Hamilton, W., Prader, A.: Steroid 17,20-desmolase deficiency: a new cause of male pseudohermaphroditism. Clin. Endocr. 1, 369-385 (1972).

49. Goebelsmann, U., Zachmann, M., Davajan, V., Israel, R., Mestman, J.H., Mishell, D.R.: Male pseudohermaphroditism consistent with 17,20-desmolase deficiency. Gynecol. Invest. 7, 138-156 (1976).

50. Forest, M.G., Lecornu, M., de Peretti, E.: Familial male pseudohermaphroditism due to 17,20-desmolase deficiency. J. clin. Endocrinol. Metab. 50, 826-833 (1980).

51. Zachmann, M., Vollmin, J.A., New, M.I., Curtius, H.C.H., Prader, A.: Congenital adrenal hyperplasia due to deficiency of 11β-hydroxylation of 17α-hydroxylated steroids. J. clin. Endocrinol. Metab. 33, 501-508 (1971).

52. Sizonenko, P.C., Riondel, A.M., Kohlberg, I.J., Paunier, L.: 11β-Hydroxylase deficiency: steroid response to sodium restriction and ACTH stimulation. J. clin. Endocrinol. Metab. 35, 281-287 (1972).

53. Gregory, T., Gardner, L.I.: Hypertensive virilizing adrenal hyperplasia with minimal impairment of synthetic route to cortisol. J. clin. Endocrinol. Metab. 43, 769-774 (1976).

54. Tan, S.Y., Noth, R.H., Mulrow, P.J.: Deoxycorticosterone and 17-ketosteroids; elevated levels in adult hypertensive patients. J.A.M.A. 240, 123-126 (1978).

55. Kuhnle, U., Chow, D., Rapaport, R., Pang, S., Levine, L.S., New, M.I.: The 21-hydroxylase activity in the glomerulosa and fasciculata of the adrenal cortex in congenital adrenal hyperplasia. J. clin. Endocrinol. Metab. 52, 534-544 (1981).

56. New, M.I., Dupont, B., Pang, S., Pollack, M., Levine, L.S.: An update of congenital adrenal hyperplasia. Recent Prog. Horm. Res. 37, 105-181 (1981).

57. Mininberg, D.T., Levine, L.S., New, M.I.: Current concepts in congenital adrenal hyperplasia. Invest. Urol. 17, 169-175 (1979).

58. Wilkins, L., Lewis, R.A., Klein, R., Rosemberg, E.: The suppression of androgen secretion by cortisone in a case of congenital adrenal hyperplasia. Bull. Johns Hopkins Hosp. 86, 249-252 (1950).

59. Bartter, F.C.: Adrenogenital syndromes from physiology to chemistry (1950-1975). In: "Congenital Adrenal Hyperplasia", Eds. Lee, P.A., Plotnick, L.P., Kowarski, A.A., Migeon, C.J., University Park Press, Baltimore, pp. 9-18 (1977).

60. Rosler, A., Levine, L.S., Schneider, B., Novogroder, M., New, M.I.: The interrelationship of sodium balance, plasma renin activity and ACTH in congenital adrenal hyperplasia. J. clin. Endocrinol. Metab. 45, 500-512 (1977).

61. Migeon, C.J., Keller, A.R., Laurence, B., Shepard, T.H.: Dehydroepiandrosterone and androsterone levels in human plasma. J. clin. Endocrinol. Metab. 17, 1051-1062 (1957).

62. Gandy, H.M., Peterson, R.E.: Measurement of testosterone and 17-ketosteroids in plasma by the double isotope dilution derivative technique. J. clin. Endocrinol. Metab. 28, 949-977 (1968).

63. Rosenfield, R.L., Eberlein, W.R.: Plasma 17-ketosteroid levels during adolescence. J. Pediat. 74, 932-936 (1969).

64. Frasier, S.D., Gafford, F., Horton, R.: Plasma androgens in childhood and adolescence. J. clin. Endocrinol. Metab. 29, 1404-1408 (1969).

65. Ito, T., Horton, R.: Dihydrotestosterone in human peripheral plasma. J. clin. Endocrinol. Metab. 31, 362-368 (1970).

66. Rosenfeld, R.S., Hellman, L., Roffwarg, H., Weitzman, E.D., Fukushima, D.K., Gallagher, T.F.: Dehydroisoandrosterone is secreted episodically and synchronously with cortisol by normal man. J. clin. Endocrinol. Metab. 33, 87-92 (1971).

67. Wieland, R.G., Chen, J.C., Zorn, E.M., Hallberg, M.C.: Correlation of growth, pubertal staging, growth hormone, gonadotropins, and testosterone levels during the pubertal growth spurt in males. J. Pediat. 79, 999-1002 (1971).

68. August, G.P., Grumbach, M.M., Kaplan, S.: Hormonal changes in puberty: III. Correlation of plasma testosterone, LH, FSH, testicular size, and bone age with male pubertal development. J. clin. Endocrinol. Metab. 34, 319-326 (1972).

69. Winter, J.S.D., Faiman, C.: Pituitary-gonadal relations in male children and adolescents. Pediatr. Res. 6, 126-135 (1972).

70. Boon, D.A., Keenan, R.E., Slaunwhite, W.R.Jr., Aceto, T.: Conjugated and unconjugated plasma androgens in normal children. Pediatr. Res. 6, 111-118 (1972).

71. Gupta, D., McCafferty, E., Rager, K.: Plasma 5α-dihydrotestosterone in adolescent males at different stages of sexual maturation. Steroids 19, 411-431 (1972).

72. Forest, M.G., Cathiard, A.M., Bertrand, J.A.: Total and unbound testosterone levels in the newborn and in normal and hypogonadal children. J. clin. Endocrinol. Metab. 36, 1132-1142 (1973).

73. Knorr, D., Bidlingmaier, F.: Diagnosis of congenital adrenal hyperplasia with 11-hydroxylase deficiency by determination of tetrahydro-11-desoxycortisol in urine. Helv. paediat. Acta (Suppl.) 34, 119-125 (1974).

74. Lee, P.A., Jaffe, R.B., Midgley, A.R.Jr.: Serum gonadotropin, testosterone and prolactin concentrations throughout puberty in boys. J. clin. Endocrinol. Metab. 39, 664-672 (1974).

75. Hopper, B.R., Yen, S.S.C.: Circulating concentrations of dehydroepiandrosterone and dehydroepiandrosterone sulfate during puberty. J. clin. Endocrinol. Metab. 40, 458-461 (1975).

76. Korth-Schutz, S., New, M.I.: Dehydroepiandrosterone sulfate levels, a rapid test for abnormal adrenal function. In: "Program and Abstracts", Annual Meeting, Lawson Wilkins

Pediatric Endocrine Society, April 1975, Denver (1975).

77. Forest, M.G., Cathiard, A.M.: Patterns of plasma testosterone and Δ^4-androstenedione in normal newborns. J. clin. Endocrinol. Metab. 41, 977-980 (1975).

78. Gupta, D., Attanasio, A., Raaf, S.: Plasma estrogen and androgen concentrations in children during adolescence. J. clin. Endocrinol. Metab. 40, 636-643 (1975).

79. Korth-Schutz, S., Levine, L.S., New, M.I.: Serum androgens in normal prepubertal and pubertal children and in children with precocious adrenarche. J. clin. Endocrinol. Metab. 42, 117-124 (1976).

80. Winter, J.S.D., Hughes, I.A., Reyes, F.I., Faiman, C.: Pituitary-gonadal relations in infancy: 2. Patterns of serum gonadal steroid concentrations in man from birth to two years of age. J. clin. Endocrinol. Metab. 42, 679-686 (1976).

81. Rivarola, M.A., Saez, J.M., Migeon, C.J.: Studies of androgens in patients with congenital adrenal hyperplasia. J. clin. Endocrinol. Metab. 27, 624-630 (1967).

82. Strott, C.A., Yoshimi, T., Lipsett, M.B.: Plasma progesterone and 17α-hydroxyprogesterone in normal men and children with congenital adrenal hyperplasia. J. clin. Invest. 48, 930-939 (1969).

83. West, C.D., Stanchfield, J.B., Atcheson, J.B., Rallison, M.L., Tyler, F.H.: Circadian variation in plasma ACTH and cortisol and its precursors in congenital adrenal hyperplasia. Clin. Res. 21, 190 (1973).

84. Lippe, B.M., LaFranchi, S.H., Lavin, N., Parlow, A., Coyotupa, J., Kaplan, S.A.: Serum 17α-hydroxyprogesterone, progesterone, estradiol, and testosterone in the diagnosis and management of congenital adrenal hyperplasia. J. Pediat. 85, 782-787 (1974).

85. Solomon, I.L., Schoen, E.J.: Blood testosterone values in patients with congenital virilizing adrenal hyperplasia. J. clin. Endocrinol. Metab. 40, 355-362 (1975).

86. Youssefnejadian, E., David, R.: Early diagnosis of congenital adrenal hyperplasia by measurement of 17-hydroxyprogesterone. Clin. Endocr. 4, 451-454 (1975).

87. Hughes, I.A., Winter, J.S.D.: The application of a serum 17-OH-progesterone radioimmunoassay to the diagnosis and management of congenital adrenal hyperplasia. J. Pediat. 88, 766-773 (1976).

88. Hughes, I.A., Winter, J.S.D.: The relationship between serum concentrations of 17-OH-progesterone and other serum and urinary steroids in patients with congenital adrenal hyperplasia. J. clin. Endocrinol. Metab. 46, 98-104 (1978).

89. Cook, D.M., Kendall, J.W., Allen, J.P., Lagerquist, L.G.: Nyctohemeral variation and suppressibility of plasma ACTH in various stages of Cushing's disease. Clin. Endocr. 5, 303-312 (1976).

90. Gupta, D., Rager, K., Klemm, W., Attanasio, A.: Diagnosis and management of congenital adrenal hyperplasia. In: "Congenital Adrenal Hyperplasia", Eds. Lee, P.A., Plotnick, L.P., Kowarski, A.A., Migeon, C.J., University Park Press, Baltimore, pp. 283-300 (1977).

91. Rappaport, R., Limal, J.M.: Linear growth and suppressive effects of hydrocortisone and 9α-fluorohydrocortisone in long-term therapy in congenital adrenal hyperplasia. In: "Congenital Adrenal Hyperplasia", Eds. Lee, P.A., Plotnick, L.P., Kowarski, A.A., Migeon, C.J., University Park Press, Baltimore, pp. 263-272 (1977).

92. Korth-Schutz, S., Virdis, R., Saenger, P., Chow, D.M., Levine, L.S., New, M.I.: Serum androgens as a continuing index of adequacy of treatment of congenital adrenal hyperplasia. J. clin. Endocrinol. Metab. 46, 452-458 (1978).

93. Pang, S., Levine, L.S., Chow, D., Faiman, C., New, M.I.: Serum androgen concentration in neonates and young infants with virilizing congenital adrenal hyperplasia. Clin. Endocr. 11, 575-584 (1979).

94. Winter, J.S.D.: Current approaches to the treatment of congenital adrenal hyperplasia. J. Pediat. 97, 81-82 (1980).

95. Childs, B., Grumbach, M.M., van Wyk, J.J.: Virilizing adrenal hyperplasia: a genetic and hormonal study. J. clin. Invest. 35, 213-222 (1956).

96. Prader, A.: Volkommen maenliche aeussere Genitalentwicklung und Salzverlust-Syndrom bei Maedchen mit kongenitalen adrenogenitalem Syndrom. Helv. paediat. Acta 13, 426-431 (1958).

97. Baulieu, E.E., Peillow, F., Migeon, C.J.: Adrenogenital Syndrome. In: "The Adrenal Cortex", Ed. Eisenstein, A.B., Little Brown, Boston, pp. 553-638 (1967).

98. Rosenbloom, A.L., Smith, D.W.: Varying expression for salt losing in related patients with congenital adrenal hyperplasia. Pediatrics 38, 215-219 (1966).

99. Muller, W., Prader, A., Kofler, J., Glatzl, J., Geir, W.: Frequency of congenital adrenal hyperplasia. Paediatr. Paedol. 14, 151-155 (1979).

100. Cohen, J.M.: Salt-losing congenital adrenal hyperplasia. Pediatrics 44, 621-622 (1969).

101. Rimoin, D.L., Schimke, R.N.: Eds. "Genetic Disorders of the Endocrine Glands", Mosby, St. Louis (1971).

102. Hirschfeld, A.G., Fleshman, J.K.: An unusually high incidence of congenital adrenal hyperplasia in the Alaskan Eskimo. J. Pediat. 75, 492-494 (1969).

103. Dupont, B., Oberfield, S.E., Smithwick, E.M., Lee, T.D., Levine, L.S.: Close genetic linkage between HLA and congenital adrenal hyperplasia. Lancet 2, 1309-1311 (1977).

104. Levine, L.S., Zachmann, M., New, M.I., Prader, A., Pollack, M.S., O'Neill, G.J., Yang, S.Y., Oberfield, S.E., Dupont, B.: Genetic mapping of the 21-hydroxylase deficiency gene within the HLA linkage group. New Engl. J. Med. 299, 911-915 (1978).

105. Dupont, B., Pollack, M.S., Levine, L.S., O'Neill, G.J., Hawkins, B., New, M.I.: Congenital adrenal hyperplasia and HLA: Joint report from the 8th International Histocompatibility Workshop. In: "Histocompatability Testing 1980", Ed. Teraski, P.I., UCLA Tissue Typing Laboratory, Los Angeles, pp. 693-706 (1980).

106. Bongiovanni, A.M.: The detection of pregnandiol and pregnantriol in the urine of patients with adrenal hyperplasia. Suppression with cortisone. Bull. Johns Hopkins Hosp. 92, 244-251 (1953).

107. Knorr, D., Bidlingmaier, F., Butenandt, O., Schnakenburg, K.V., Wagner, W.: Test for heterozygosity of congenital adrenal hyperplasia. In: "Congenital Adrenal Hyperplasia", Eds. Lee, P.A., Plotnick, L.P., Kowarski, A.A., Migeon, C.J., University Park Press, Baltimore, pp. 495-500 (1977).

108. Lorenzen, F., Pang, S., New, M.I., Pollack, M.S., Oberfield S.E., Dupont, B., Chow, D., Levine, L.S.: Studies of the C-21 and C-19 steroids and HLA genotyping in siblings and parents of patients with congenital adrenal hyperplasia due to 21-hydroxylase deficiency. J. clin. Endocrinol. Metab. 50, 572-579 (1980).

109. Lorenzen, F., Pang, S., New, M.I., Dupont, B., Pollack, M.S., Chow, D., Levine, L.S.: Hormonal phenotype and HLA-genotype in families of patients with congenital adrenal hyperplasia. Pediatr. Res. 13, 1356-1370 (1979).

110. Grosse-Wilde, H.J., Weil, J., Albert, E., Scholz, S., Bidlingmaier, F., Sippel, W.G., Knorr, D.: Genetic linkage studies between congenital adrenal hyperplasia and the HLA blood group system. Immunogenetics 8, 41-49 (1979).

111. Lejeune-Lenain, C., Cantraine, F., Dufrasnes, M., Prevot, F., Wolter, R., Franckson, J.R.M.: An improved method for the detection of heterozygosity of congenital virilizing adrenal hyperplasia. Clin. Endocr. 12, 525-535 (1980).

112. Decourt,M.J., Jayle, M.F., Baulieu, E.: Virilisme cliniquement tardif avec excretion de pregnanetriol et insuffisance de la production du cortisol. Ann. Endocrinol. 18, 416-422 (1957).

113. Newmark, S., Dluhy, R., Williams, G., Pochi, P., Rose, L.: Partial 11 and 21-hydroxylase deficiencies in hirsute women. Am. J. Obstet. Gynecol. 127, 594-598 (1977).

114. Rosenwaks, Z., Lee, P.A., Jones, G.S., Migeon, C.J., Wentz, A.C.: An attenuated form of congenital adrenal hyperplasia. J. clin. Endocrinol. Metab. 49, 335-339 (1979).

115. Bouchard, P., Kuttenn, F., Mowszowicz, I., Schaison, G., Raux-Eurin, M.-C., Mauvais-Jarvis, P.: Congenital adrenal hyperplasia due to partial 21-hydroxylase deficiency. Acta Endocrinol. 96, 107-111 (1981).

116. New, M.I., Lorenzen, F., Pang, S., Gunczler, P., Dupont, B., Pollack, M., Levine, L.S.: "Acquired" adrenal hyperplasia with 21-hydroxylase deficiency is not the same genetic disorder as congenital adrenal hyperplasia. J. clin. Endocrinol. Metab. 48, 352-359 (1979).

117. Morillo, E., Gardner, L.I.: Genetics of acquired and congenital adrenal hyperplasia. Lancet 2, 202-203 (1979).

118. Laron, Z., Pollack, M.S., Zamir, R., Roitman, A., Dickerman, Z., Levine, L.S., Lorenzen, F., O'Neill, G.J., Pang, S., New, M.I., Dupont, B.: Late onset 21-hydroxylase deficiency and HLA in the Ashkenazi population: a new allele at the 21-hydroxylase locus. Human Immunol. 1, 55-66 (1980).

119. Blankstein, J., Faiman, C., Reyes, F.I., Schroeder, M.L., Winter, J.S.D.: Adult-onset familial adrenal 21-hydroxylase deficiency. Am. J. Med. 68, 441-448 (1980).

120. Migeon, C.J., Rosenwaks, Z., Lee, P.A., Urban, M.D., Bias, W.B.: The attenuated form of congenital adrenal hyperplasia as an allelic form of 21-hydroxylase deficiency. J. clin. Endocrinol. Metab. 51, 647-649 (1980).

121. Pollack, M.S., Levine, L.S., O'Neill, G.J., Pang, S., Lorenzen, F., Kohn, B., Randanini, G.F., Chiumello, G., New, M.I., Dupont, B.: HLA linkage and B14, DR1, Bfs haplotype association with the genes for late onset and cryptic 21-hydroxylase deficiency. Am. J. Human Genet. 33, 540-550 (1981).

122. Levine, L.S., Dupont, B., Lorenzen, F., Pang, S., Pollack, M., Oberfield, S., Kohn, B., Lerner, A., Cacciari, E., Mantero, F., Cassio, A., Scaroni, C., Chiumello, G., Rondanini, G.F., Gargantini, L., Giovannelli, G., Virdis, R., Bartolotta, E., Migliori, C., Pintor, C., Tato, L., Barboni, F., New, M.I.: Cryptic 21-hydroxylase deficiency in families of patients with classical congenital adrenal hyperplasia. J. clin. Endocrinol. Metab. 51, 1316-1324 (1980).

123. Zachmann, M., Prader, A.: Unusual heterozygotes of congenital adrenal hyperplasia due to 21-hydroxylase deficiency. Acta Endocrinol. 87, 557-565 (1978).

124. Zachmann, M., Prader, A.: Unusual heterozygotes of congenital adrenal hyperplasia due to 21-hydroxylase deficiency confirmed by HLA tissue typing. Acta Endocrinol. 95, 542-546 (1979).

125. McKusick, V.A.: Phenotypic diversity of human diseases resulting from allelic series. Am. J. Human Genet. 25, 446-456 (1973).

126. Klouda, P.T., Harris, R., Price, D.A.: HLA and congenital adrenal hyperplasia. Lancet 2, 1046 (1978).

127. Pollack, M.S., Levine, L.S., Zachmann, M., Prader, A., New, M.I., Oberfield, S., Dupont, B.: Possible genetic linkage disequilibrium between HLA and the 21-hydroxylase deficiency gene (congenital adrenal hyperplasia). Transplant Proc. XI, 1315-1316 (1979).

128. Pollack, M.S., Levine, L.S., O'Neill, G.J., Pang, S., Lorenzen, F., Kohn, B., Rondanini, G.F., Chiumello, G., New, M.I., Dupont, B.: HLA Linkage and B14, DR1, BfS haplotype association with the genes for late onset and cryptic 21-hydroxylase deficiency. Am. J. Human Genet. 58, 331-337 (1981).

129. Mantero, F., Scaroni, C., Pasini, C., Fagliolo, U.: No linkage between HLA and congenital adrenal hyperplasia due to 17-alpha-hydroxylase deficiency. New Engl. J. Med. 303, 530 (1980).

130. Pang, S., Levine, L.S., Lorenzen, F., Chow, D., Pollack, M.S., Dupont, B., Genel, M., New, M.I.: Hormonal studies in obligate heterozygotes and siblings of patients with 11β-hydroxylase deficiency congenital adrenal hyperplasia. J. clin. Endocrinol. Metab. 50, 586-589 (1980).

131. Brautbar, C., Theodor, R., Sack, J., Levine, C., Dupont, B., Levine, L.S., Sharon, R., Smaller, S., Cohen, T., Rosler, A.: HLA in a selective aldosterone biosynthetic defect due to type 2 corticosterone methyl-oxidase deficiency. Tissue Antigens 17, 212-216 (1981).

132. Pang, S., Hotchkiss, J., Drash, A.L., Levine, L.S., New, M.I.: Microfilter paper method for 17-α-progesterone radioimmunoassay: its application for rapid screening for congenital adrenal hyperplasia. J. clin. Endocrinol. Metab. 45, 1003-1008 (1977).

133. Werder, E.A., Siebenmann, R.E., Knorr-Murset, G., Zimmermann, A., Sizonenko, P.C., Theintz, P., Girard, J., Zachmann, M., Prader, A.: The incidence of congenital adrenal hyperplasia in Switzerland: a survey of patients born in 1960 to 1974. Helv. paediat. Acta 35, 5-11 (1980).

134. Price, D.A., Klouda, P.T., Harris, R.: HLA and congenital adrenal hyperplasia linkage confirmed. Lancet 1, 930-931 (1978).

135. Pang, S., Murphey, W., Levine, L.S., Spence, D.A., Leon, A., LaFranchi, S., Surve, A.S., New, M.I.: A pilot newborn secreening for congenital adrenal hyperplasia at the New York Hospital and Alaska. J. clin. Endocrinol. Metab. 55, 413-420 (1982).

136. LaFranchi, S.H., Murphey, W.H., Foley, T.P.Jr., Larsen, P.R., Buist, N.R.M.: Neonatal hypothyroidism detected by the northwest regional screening program. Pediatrics 63, 180-191 (1979).

137. Natoli, G., Moschini, L., Acconcia, P., Albino, G., Costa, P., Pansa, G.: Newborn screening for 21-hydroxylase deficiency: Radioimmunoassay of 17-α-hydroxyprogesterone by microfilter paper method. In: "Recent Progress in Pediatric Endocrinology", Serono Series, Eds. Chiumello, G., Sperling, M., Raven Press, New York, pp. 285-290 (1983).

138. Cacciari, E., Balsamo, A., Cassio, A., Piazza, S., Bernardi, F., Salardi, S., Cicognani, A., Pirazzili, P., Zappulla, F., Capelli, M., Paolini, M.: Neonatal screening for congenital adrenal hyperplasia. Lancet 1, 1069 (1982).

139. Jeffcoate, T.N.A., Fliegner, J.R.H., Russell, S.H., Davis, J.C., Wade, A.P.: Diagnosis of the adrenogenital syndrome before birth. Lancet 2, 553-555 (1965).

140. Frasier, S.D., Thorneycroft, I.H., Weiss, B.A., Horton, R.: Elevated amniotic fluid concentration of 17α-hydroxyprogesterone in congenital adrenal hyperplasia. J. Pediat. 86, 310-312 (1975).

141. Nagamani, M., McDonough, P.G., Ellegood, J.O., Mahesh, V.B.: Maternal and amniotic fluid 17α-hydroxyprogesterone levels during pregnancy: diagnosis of congenital adrenal hyperplasia in utero. Am. J. Obstet. Gynecol. 130, 791-794 (1978).

142. Hughes, I.A., Laurence, K.M.: Antenatal diagnosis of congenital adrenal hyperplasia. Lancet 2, 7-9 (1979).

143. Pang, S., Levine, L.S., Cederqvist, L.L., Fuentes, M., Riccardi, V.M., Holcombe, J.H., Nitowsky, H.M., Sachs, G., Anderson, C.E., Duchon, M.A., Owens, R., Merkatz, I., New, M.I.: Amniotic fluid concentrations of Δ^5 and Δ^4 steroids in fetuses with congenital adrenal hyperplasia due to 21-hydroxylase deficiency and in anencephalic fetuses. J. clin. Endocrinol. Metab. 51, 223-229 (1980).

144. Pollack, M.S., Levine, L.S., Pang, S., Owens, R.P., Nitowsky, H.M., Maurer, D., New, M.I., Duchon, M., Merkatz, I.R., Sachs, G., Dupont, B.: Prenatal diagnosis of congenital adrenal hyperplasia by HLA typing. Lancet 1, 1107-1108 (1979).

145. Pollack, M.S., Heagney, S.D., Braun, D.Jr., O'Neill, G.J.: Technical and theoretical considerations in the HLA typing of amniotic fluid cells for prenatal diagnosis and paternity testing. Prenatal Diagnosis 1, 183-195 (1981).

146. Rosler, A., Leiberman, E., Rosenmann, A., Ben-Uzilio, R., Weidenfeld, J.: Prenatal diagnosis of 11β-hydroxylase deficiency congenital adrenal hyperplasia. J. clin. Endocrinol. Metab. 49, 546-551 (1979).

147. Schumert, Z., Rosenmann, A., Landau, H., Rosler, A.: 11-Deoxycortisol in amniotic fluid: prenatal diagnosis of congenital adrenal hyperplasia due to 11β-hydroxylase deficiency. Clin. Endocr. 12, 257-260 (1980).

148. Nihoul-Fekete, C.: Feminizing genitoplasty in the intersex child. In: "The Intersex Child", Ed. Josso, N., vol.8 of "Pediatric Adolescent Endocrinology", Ed. Laron, Z., Karger, Basel, pp. 247-260 (1981).

149. Hubble, D.: Congenital adrenal hyperplasia. In: "Basic Concepts of Inborn Errors and Defects of Steroid Biosynthesis", Eds. Holt, K.S., Raine, D.N., Livingstone, Edinburgh, pp. 68-74 (1971).

150. Qazi, Q.H., Thompson, M.W.: Incidence of salt-losing form of congenital virilizing adrenal hyperplasia. Arch. Dis. Childh. 47, 302-304 (1972).

151. Mauthe, I., Laspe Knorr, D.: Zur Haufigkeit des Kongenitalen Adrenogenitalen Syndroms, München 1963-1972. Klin. Pediat. 189, 172-176 (1980).

152. New, M.I., Peterson, R.E.: Disorders of aldosterone secretion in childhood. Pediatric Clinics of North America 13, 43-58 (1966).

153. New, M.I., Dupont, B., Grumbach, K., Levine, L.S.: Congenital adrenal hyperplasia and related conditions. In: "The Metabolic Basis of Inherited Disease", Eds. Stanbury, J.B., Wyngaarden, J.B., Fredrickson, D.S., Goldstein, J.L., Brown, M.S., McGraw-Hill, New York, pp. 973-1000 (1981).

STEROIDOGENESIS IN THE NORMAL AND ABNORMAL HUMAN TESTIS

J. P. Preslock
Department of Obstetrics, Gynecology and Reproductive Sciences,
University of Texas Medical School, Houston, Texas 77025, U.S.A.

Introduction

It has been demonstrated both *in vivo* and *in vitro* that the testis of mammalian species including the human can enzymatically convert a number of specific precursors into androgens. Precursors which have been demonstrated to be converted include acetate (1), cholesterol (2), pregnenolone (3) and progesterone (4). The primary androgen formed from these precursors is testosterone and the principal aim of many of these investigations was to determine the mechanism by which the testis converted these precursors into testosterone.

The synthesis of testosterone occurs primarily within the interstitial cells of Leydig, which are specialized cells within the angular spaces among the seminiferous tubules. Some testosterone synthesis occurs within the Sertoli cells and within the germinal cells of the seminiferous tubules (5). The Leydig cell is characterized by an extensive smooth endoplasmic reticulum and a well-developed Golgi apparatus, along with numerous lipid droplets and lysosomes, structures which are indicative of actively secreting steroidogenic cells (6). The mitochondria and the smooth endoplasmic reticulum contains the vast enzymatic apparatus necessary for the biosynthesis of testosterone within the Leydig cell.

The synthesis of steroid molecules including testosterone involves a complex series of reactions in which acetate moieties are linked together to form a series of intermediate ste-

Steroid Converting Enzymes and Diseases

rols which include mevalonic acid, farnesyl pyrophosphate, squalene and lanosterol, the latter being the immediate precursor of cholesterol. Many studies have shown that the rate-limiting step for the biosynthesis of steroids by the testis is the conversion of cholesterol to pregnenolone by the cholesterol side-chain cleavage enzyme complex present within the mitochondria of the Leydig cell (7, 8). This critical conversion of cholesterol to pregnenolone has been proposed to be the result of the activities of a number of testicular enzymes (9), and this enzymic conversion is stimulated by LH from the pituitary gland (10).

Testicular Steroidogenic Enzymes and the Conversion of Pregnenolone to Testosterone

In mammalian species including the human the conversion of pregnenolone to testosterone is the result of a complex series of reactions which are catalyzed by specific testicular enzymes. Pregnenolone is converted into testosterone through either the Δ^4 (4-ene) or the Δ^5 (5-ene) pathway. Conversion through the 4-ene pathway involves the following conversions: pregnenolone → progesterone → 17-hydroxyprogesterone → androstenedione → testosterone and through the 5-ene pathway, the following conversions: pregnenolone → 17-hydroxypregnenolone → dehydroepiandrosterone → androstenedione → testosterone. Alternate pathways which may also be utilized as part of the 5-ene pathway are dehydroepiandrosterone → 5-androstenediol → testosterone, and 17-hydroxypregnenolone → 17-hydroxyprogesterone → androstenedione → testosterone. Axelrod (11) and Yanaihara and Troen (12) have reported that the human testis utilizes the 5-ene pathway.

In the human testis there is a series of enzymes which react with specific sites upon the steroid molecules and bring about enzymic conversions which result in the formation of specific steroid metabolites. Although there is an obvious degree of specificity to each of these conversions, there is also a degree of non-specificity since a given enzyme complex may catalyze the conversion of more than one steroid substrate into their respective metabolic products. For instance, 3β-hydroxysteroid: NAD (P) oxidoreductase (3β-hydroxysteroid dehydrogenase; 1.1.1.51) and 3-ketosteroid-5-ene⟶ 4-ene-isomerase (isomerase), dehydrogenates the 3β-hydroxy group and isomerizes the 5-ene bond of pregnenolone to form progesterone. This same enzyme complex converts 17-hydroxypregnenolone to 17-hydroxyprogesterone, dehydroepiandrosterone to androstenedione and 5-androstenediol to testosterone. Many of these reactions are non-competitively inhibited by the steroids formed as metabolic products of these reactions. For example, the enzymic conversion of pregnenolone to progesterone is inhibited by progesterone, 17-hydroxyprogesterone, androstenedione and testosterone.

The enzyme steroid 17-hydroxylase (17-hydroxylase; 1.99.1.9) converts pregnenolone and progesterone into their respective 17-hydroxylated products, with NADPH as a necessary cofactor (13). The 17-hydroxylases are mixed function oxidases which involve NADPH-cytochrome P-450 reductase, a flavoprotein, and an unknown factor which may be a phospholipid and which transfers electrons from NADPH to cytochrome P-450. Whether the 17-hydroxylation of pregnenolone and progesterone results from the activity of a single enzyme or separate enzymes has not been conclusively determined in the human. However, in rat testis, Tamaoki (14) reported that progesterone was the primary substrate for 17-hydroxylase, while Fevold and Drummond (15) have reported pregnenolone as the primary substrate. Kremers (16) showed that 17-hydroxylase in the testis is a single enzyme which can 17-hydroxylate both pregnenolone and progesterone, but with a greater affinity for pregnenolone.

The enzyme 17-hydroxysteroid C(17)-C(20)-lyase may also act upon more than one substrate in the human testis. This enzyme side-chain cleaves the side-chain of 17-hydroxyprogesterone to androstenedione and that of 17-hydroxypregnenolone to dehydroepiandrosterone. NADPH is required as a cofactor, as is molecular oxygen (17). As with 17-hydroxylase, it is yet unclear whether the conversion of 4-ene and 5-ene steroids by C(17)-C(20)-lyase involve the same or separate enzymes.

The reduction of androstenedione to testosterone and dehydroepiandrosterone to 5-androstenediol is accomplished by 17β-hydroxysteroid: NAD(P) oxidoreductase (17β-hydroxysteroid dehydrogenase: 1.1.1.51) and NADPH is required. Oshima et al. (18) reported that 17β-hydroxysteroid dehydrogenase more readily converted androstenedione to testosterone than dehydroepiandrosterone to 5-androstenediol. This enzyme is unique among testicular steroidogenic enzymes since it can also oxidize testosterone into androstenedione provided that the oxidizing cofactor NADP+ is included in the reaction mixture (19).

Several other enzymes have also been identified in the human testis. 20α-Hydroxysteroid dehydrogenase converts pregnenolone to 20α-dihydropregnenolone, progesterone to 20α-dihydroprogesterone, 17-hydroxyprogesterone to 17-hydroxy-20α-dihydroprogesterone (17α,20α-dihydroxyprogesterone), and 17-hydroxypregnenolone to 17-hydroxy-20α-dihydropregnenolone (17α, 20α-dihydroxypregnenolone). Similarly, 20β-hydroxysteroid dehydrogenase can convert pregnenolone, progesterone, 17-hydroxyprogesterone and 17-hydroxypregnenolone to their respective 20β-hydroxylated derivatives. 16α-Hydroxylase can convert a number of substrates including progesterone to 16α-hydroxylated derivatives.

Oshima et al. (20) and Inano and Tamaoki (21) have postulated that 20α and 20β-hydroxysteroid dehydrogenases may serve as intracellular regulating enzymes for testosterone synthesis. According to their hypothesis, the conversion of pregnenolones and progesterones to 20α and 20β-hydroxylated derivatives removes these steroids from the substrate pool and furthermore, the products formed from the activities of these enzymes may competitively or noncompetitively inhibit the activities of enzymes necessary for the synthesis of testosterone such as

17-hydroxylase, and the lyase.

5α-Reductase (4-ene-5α-reductase) reduces the double bond between C-4 and C-5 of a number of substrates with the formation of 5α-reduced androgens. The best known of these reactions is the conversion of testosterone to 5α-dihydrotestosterone (5α-DHT). 5α-DHT can be further metabolised to other 5α-reduced steroids such as androsterone, isoandrosterone and 5α-androstanediol.

"Aromatase" refers to a series of enzymes which by a specific sequence of events converts testosterone and androstenedione to estrogens. For example, the aromatization of testosterone by "aromatase" involves the conversion of testosterone to 19-hydroxy-testosterone, to 19-oxo-testosterone, to 19-carboxy-testosterone, to 19-nor-testosterone (CO_2 is a by-product) and finally to estradiol. Through a similar sequence of steps androstenedione can be converted into estrone. 17β-Hydroxysteroid dehydrogenase can interconvert estrone and estradiol within the human testis.

Androgens Secreted from the Human Testis

In humans testosterone is the major androgen synthesized by the testis. The human testis secretes from 5 to 12 mg testosterone per day together with smaller amounts of other androgens, dehydroepiandrosterone, androstenedione, 5-androstenediol and dihydrotestosterone (22). Testosterone in the blood is bound with a high affinity and a low capacity to testosterone-estrogen binding globulin (TeBG) and with a low affinity but a high capacity to albumin. Due to this protein binding, only about 3% of plasma testosterone is unbound and therefore biologically active. This amount of free plasma testosterone can be varied by factors which either increase or decrease the concentration of these serum binding proteins. There is a nocturnal elevation in serum testosterone levels which is not correlated with any variation in the serum levels of LH, FSH and prolactin (23, 24).

Berthold (25) in 1849 provided the first evidence of hormone production by the human testis but it was not until 1934 that Ruzicka (26) identified testosterone in the testes. Savard et al. (27) provided the first conclusive evidence for the secretion of androgens from the human testis by perfusing it with ^{14}C-acetate and identifying labelled testosterone and androstenedione in the perfusate.

Site of Androgen Production in the Human Testis

Tamaoki et al. (28) showed that the major site of androgen production in the human testis was the Leydig cells, although the seminiferous tubules are also capable of converting pregnenolone to testosterone (29). The Leydig cells are localized as clusters within the interstitial tissue interspersed between the seminiferous tubules. In the human testis, the Leydig cells are characterized as large round cells with round or oval nuclei (30), with a well-developed smooth endoplasmic reticulum (31). The mitochondria are well-developed with numerous villous cristae which van der Vusse et al. (32) have proposed as the site of pregnenolone synthesis from cholesterol. The smooth endoplasmic reticulum has been reported to be the site at which pregnenolone is converted into testosterone since the enzymes necessary for the conversion steps are localized within this subcellular organelle (33).

Target-Tissue Metabolism of Testosterone

In vitro studies have demonstrated that several androgen-responsive tissues contain 5α-reductase which reduces testosterone to 5α-dihydrotestosterone (34) and this steroid has been proposed as the active form of testosterone in testosterone-dependent target tissues (35, 36).

Testicular Steroidogenesis in the Normal Human Testis

Information concerning the *in vitro* biosynthesis of androgens in the normal human testis was initially derived with testicular tissue obtained at orchiectomy from elderly patients who had carcinoma of the prostate and who were usually undergoing estrogen therapy (37-39) or from patients with testicular feminization (40-42). Neither of these sources provided what could be considered normal healthy adult human testicular tissue. However, investigations with normal human testicular tissue have demonstrated that the 5-ene pathway is the predominant one for the synthesis of testosterone (11, 12, 40). A number of reports nevertheless have demonstrated that the human testis can utilise both the 4-ene and the 5-ene pathways (43, 44). Steinberger et al. (45) reported that progesterone could be metabolised by either 17-hydroxylase to form 17-hydroxyprogesterone or by 20α-hydroxysteroid dehydrogenase to form 20α-dihydroprogesterone with the former pathway predominant. The predominant pathway for the metabolism of 17-hydroxyprogesterone was to androstenedione and it was also metabolised to 17α-hydroxy-20α-dihydroprogesterone and 17, 20β-dihydroprogesterone. Aromatase was also active since estrone and estradiol were also identified within the incubates.

Yanaihara and Troen (12) reported that the 5-ene intermediates 17-hydroxypregnenolone and dehydroepiandrosterone were more efficiently converted to androgens than were the 4-ene intermediates progesterone and 17-hydroxyprogesterone. They further reported that the metabolites formed from progesterone changed with increasing times of incubation since, although 17-hydroxyprogesterone was the major metabolite formed from progesterone by 17-hydroxylase during the early portion of the incubations, testosterone became the predominant metabolite formed later in the incubations. Sulcova and Starka (46) demonstrated that 17β-hydroxysteroid dehydrogenase could efficiently convert the dehydroepiandrosterone to 5-androstenediol but that 3β-hydroxysteroid dehydrogenase and isomerase were inefficient in converting dehydroepiandrosterone to androstenedione.

Kjessler and Berg (47) investigated the *in vitro* metabolism of progesterone by testicular tissue from nine apparently healthy adult human males. As in previous work (45), progesterone was metabolised by both 17-hydroxylase and 20α-hydroxysteroid dehydrogenase to form 17-hydroxyprogesterone and 20α-dihydroprogesterone, which accounted for 17-47% and 13-49% respectively, of the metabolites. Since testosterone and androstenedione were only minor metabolites, it was suggested that the endogenous pool of testosterone and androstenedione may have interfered with the conversion of 17-hydroxyprogesterone into androstenedione, thus suggesting a feedback-inhibition of androgen synthesis by endogenous androgens.

Rosner and Macombe (48) incubated normal human testes from seven patients with radiolabelled pregnenolone and dehydroepiandrosterone and reported these substrates were converted to 5-androstenediol. Furthermore, when equal amounts of ^{3}H-dehydroepiandrosterone and ^{14}C-17-hydroxyprogesterone were incubated with normal human testes, the $^{3}H/^{14}C$ ratio was higher in testosterone than androstenedione, suggesting a pathway to testosterone which bypassed androstenedione. 5-Androstenediol was efficiently converted to testosterone, suggesting that the 5-ene pathway for testosterone formation in human testes may involve the conversion of dehydroepiandrosterone to 5-androstenediol and then to testosterone.

In summary these results suggest that, although the human testis converts pregnenolone to testosterone predominantly through the 5-ene pathway due possibly to the efficient conversion of pregnenolone to 17-hydroxypregnenolone, intermediates of both the 5-ene and 4-ene pathways can be converted to testosterone. Furthermore, the 5-ene pathway of testosterone synthesis may not involve androstenedione as an obligatory intermediate.

Testicular Steroidogenesis in the Abnormal Human Testis

For the purposes of our discussion, this section will investigate the current state of knowledge with regard to the testicular synthesis of androgens in a number of syndromes in which normal gonadal function is known to be disrupted. In general there is little information available for testicular steroidogenesis.

a. The infertile male

In a recent study, Rodriguez-Rigau et al. (49) obtained testicular biopsies from 33 infertile patients and from a young healthy fertile volunteer. The tissue was incubated with radioactive pregnenolone and the major metabolites were isolated and identified. In eight of the 33 infertility patients, there were substantial differences from the remaining 25 and from the normal male. In these eight patients the activities of 17-hydroxylase, lyase and 20α- and 20β-hydroxysteroid dehydrogenase were high since 17-hydroxypregnenolone, dehydroepiandrosterone, 20α-dihydropregnenolone and 20α-dihydro-17-hydroxypregnenolone were major metabolites and there was only a minute conversion to progesterone, 17-hydroxyprogesterone, androstenedione and testosterone. In the remaining 25 infertile patients and in the normal male pregnenolone was efficiently converted to testosterone and androstenedione through the 5-ene pathway. These authors speculated that in the eight patients in which the conversion of pregnenolone to testosterone and androstenedione was low, there may have been an impairment in 3β-hydroxysteroid dehydrogenase and isomerase activities which resulted in a decreased conversion of dehydroepiandrosterone to androstenedione and a decreased metabolism of pregnenolone to progesterone. The decreased activities of these enzymes was correlated with impaired spermatogenic function in the infertile patients (49).

b. Klinefelter's syndrome

The pattern of testicular steroidogenesis is impaired in patients with Klinefelter's syndrome compared to normal males. Steinberger et al. (45) reported that the normal human testis efficiently converted progesterone to 17-hydroxyprogesterone, 20α-dihydroprogesterone, androstenedione and testosterone. However, testicular tissue from a patient with Klinefelter's syndrome converted progesterone primarily to 17-hydroxyprogesterone with only a very low conversion to testosterone and androstenedione. These results suggest that in Klinefelter's patients there may be an impairment of lyase activity. Sharma et al. (50) reported that the ratio of estrone and estradiol produced from 17-hydroxyprogesterone *in vitro* relative to testosterone was higher in testicular tissue of patients with Klinefelter's syndrome than in normal males. They suggested that the decreased testosterone production in the Klinefelter's patients may have been due to increased aromatase activity, resulting in an increase conversion of testosterone into estrogens. Lipsett et al. (51) also reported a low *in vivo* production rate of testosterone in Klinefelter's patients. The steroidogenic disorders in the testis of Klinefelter's patients may reflect impaired enzymic activity within the Leydig cells of these patients (52).

c. Testicular feminization

Bell (41) investigated the steroid metabolic pathways for androgen production in testicular tissue obtained from patients with "complete" and "incomplete" forms of testicular feminization and reported substantial differences in the metabolism of specific precursors. Testicular tissue from both types were incubated with radioactive pregnenolone, 17-hydroxypregnenolone or dehydroepiandrosterone. With testicular tissue from patients with the "incomplete" form, there was a substantially greater conversion of pregnenolone to progesterone, of 17-hydroxypregnenolone to 17-hydroxyprogesterone, and of dehydroepiandrosterone to androstenedione. Since 3β-hydroxysteroid dehydrogenase and isomerase catalyzed these conversions, Bell (41) suggested that there was a relatively greater activity of the 3β-hydroxysteroid dehydrogenase-isomerase system in the "incomplete" form of testicular feminization compared to the "complete" form. He also demonstrated that in seven patients with testicular feminization, the pathway for testosterone production was identical to that of normal individuals in that the 5-ene pathway was predominant. However, in these feminized patients, there was an accumulation of dehydroepiandrosterone and androstenedione which did not occur when testicular tissue from normal individuals was used, suggesting decreased lyase and 17β-hydroxysteroid dehydrogenase activities in feminized patients. Conversely, Schindler (53) reported increased 17-hydroxylase, lyase and 17β-hydroxysteroid dehydrogenase activities in testicular tissue obtained from patients with testicular feminization compared to normal controls.

d. Prostatic carcinoma

Slaunwhite et al. (54), using testicular tissue obtained from patients with prostatic carcinoma who had been treated with estradiol for extended periods of time, reported that exogenous estradiol prevented the 17β-reduction of androstenedione to testosterone but did not affect the conversion of 17-hydroxyprogesterone to androstenedione, suggesting an inhibition of 17β-hydroxysteroid dehydrogenase by exogenous estradiol in these prostatic patients. In a similar study, Schoen (44) confirmed these results and suggested the effect of estrogens may be through a suppression of gonadotropin secretion. Tamaoki and Shikita (55) however reported that treatment of a prostatic carcinoma patient with estrogen resulted in a suppression of testicular microsomal 17-hydroxylase and lyase activities with a slight increase in the activity of 17β-hydroxysteroid dehydrogenase.

Testicular Steroidogenesis in Normal Patients Undergoing Estrogen Therapy

In addition to the above studies utilizing testicular tissue from adult males undergoing estrogen therapy for prostatic carcinoma, Rodriguez-Rigau et al. (56) conducted an extensive series of *in vitro* investigations with testicular tissue from adult male transsexuals prior to and after estrogen therapy. Prior to estrogen treatment, the major metabolite formed from progesterone was 17-hydroxyprogesterone; 17-hydroxyprogesterone, androstenedione and testosterone comprised 54% of the metabolites and 20α-dihydroprogesterone a further 2.7%. With tissue obtained after estrogen administration, there was a marked inhibition of progesterone metabolism since more than 70% of the substrate was not metabolised. The major metabolite formed in these patients was 20α-dihydroprogesterone (5.3-13.2%), while testosterone, androstenedione and 17-hydroxyprogesterone comprised less than 2% of the metabolites. These studies suggest that exogenous estrogen blocks 17-hydroxylase activity but enhances the conversion of progesterone to 20α-dihydroprogesterone. In a subsequent study, Rodriguez-Rigau et al. (57) reported that in the testis of a 28-year-old transsexual male treated with estrogen the metabolism of cholesterol, pregnenolone and progesterone to testosterone were also blocked. However, androstenedione substrate was nearly completely (95.1%) metabolised to testosterone. When testosterone was the substrate, only 18% was converted to androstenedione, 5α-androstanediol, dihydrotestosterone and estradiol, the remainder being unmetabolised. These studies suggest that exogenous estrogen blocks the formation of testosterone by inhibiting the activities of 3β-hydroxysteroid dehydrogenase and isomerase, 17-hydroxylase or lyase, but does not inhibit the 17β-reduction of androstenedione to testosterone.

Effects of Age upon Testicular Steroidogenesis in the Human

A number of studies have suggested that the steroid metabolic pattern of the human testis depends upon the ages of the individuals from whom the testicular tissue is obtained. Axelrod (11) noted substantial differences in the metabolism of pregnenolone in a 60-year-old man compared to a 16-year-old boy. In these studies, testicular tissue from the 60-year-old man converted pregnenolone primarily to 17-hydroxyprogesterone, with a relatively low conversion to androstenedione and testosterone, suggesting a possible impairment in the activity of lyase. Testicular tissue from the 16-year-old boy converted pregnenolone primarily to testosterone, with only a small portion recovered as 17-hydroxyprogesterone. With 17-hydroxyprogesterone as the substrate, testicular tissue from the 60-year-old man converted a substantial portion to 17-hydroxy-20α-dihydroprogesterone with a relatively low conversion to testosterone.

There are other reports which suggest impaired lyase activity with increasing age. Steinberger et al. (39, 58) reported that only about 50% of progesterone was metabolised by testicular tissue obtained from a 71-year-old man with prostatic carcinoma not treated with estrogen with a substantial majority of the radioactivity recovered as 17-hydroxyprogesterone and similar results were obtained by Bell and Lacy (40). However, tissue from a younger male converted progesterone primarily to testosterone.

There may also be age-dependent differences in the activity of human testicular 5α-reductase. The prepubertal human testis has little 5α-reductase activity but with increasing age its activity has been reported to increase (59-61).

Berg et al. (62) investigated the steroid metabolic pathways in young males ranging in age from 5 to 15 years, using progesterone as substrate. The metabolic products identified (20α-dihydroprogesterone, 17-hydroxyprogesterone, androstenedione and testosterone, with no appreciable 5α-reduced metabolites) were identical to those of adult males described previously (45, 47) although the prepubertal and adolescent samples produced a relatively larger proportion of 20α-dihydroprogesterone and 17-hydroxyprogesterone. In prepubertal boys below the age of 11, 20α-dihydroprogesterone accounted for 45 to 82% of the progesterone metabolites.

In studies with adult testicular tissue, there was only a slight reduction of testosterone and androstenedione to 5α-metabolites (46) although in a similar study, a higher 5α-reductase activity in a 72-year-old man compared to a younger male was reported (40).

In a subsequent report, Sharma et al. (63) incubated testicular tissue obtained from a young post-pubertal male with radioactive progesterone or 17-hydroxyprogesterone. With progesterone as substrate, 17-hydroxyprogesterone was the major metabolite formed, while testosterone and androstenedione were minor metabolites. With 17-hydroxyprogesterone as substrate, the majority of substrate remained unconverted, but there was a slightly higher percentage of conversion into testosterone and androstenedione than when progesterone was the substrate. Other metabolites identified included 20α-dihydroprogesterone and 17-hydroxy-20α-dihydroprogesterone, but with no 5α-reduced metabolites. This failure to efficiently convert progesterone and 17-hydroxyprogesterone into testosterone suggests a deficiency in either 17-hydroxylase or lyase activity in younger males, a situation similar to that of the estrogen-treated adult male as reported by Rodriguez-Rigau et al. (56).

In summary, the results of these studies suggest that lyase activity is low in prepubertal and pubertal boys, is elevated post-pubertally and then decreases with increasing age of the individual. 5α-Reductase activity also appears to be low in prepubertal and pubertal boys but may increase with age.

Testicular Enzymes and Puberty

Testicular enzymes have been implicated in the initiation of puberty in the human.

In the human testis, the acquisition of the somatic conversion indicative of puberty are dependent upon the attainment of sufficiently high levels of serum testosterone. The attainment of these high levels of serum testosterone are in part regulated by the activities of the testicular enzymes 17-hydroxylase and 20α-hydroxysteroid dehydrogenase, which results in the conversion of pregnenolone or progesterone into either 17-hydroxylated intermediates which can be converted into testosterone or into 20α-hydroxylated intermediates which cannot.

Fan et al. (64) suggested that the testicular synthesis of androgens in the human may be regulated by 20α-hydroxysteroid dehydrogenase which competes with 17-hydroxylase for pregnenolone and progesterone or which competitively inhibits the activity of 17-hydroxylase by forming 20α-dihydropregnenolone and 20α-dihydroprogesterone from pregnenolone and progesterone. In these studies, 20α-dihydropregnenolone and 20α-dihydroprogesterone competitively inhibited the 17-hydroxylation of pregnenolone and progesterone, while 17-hydroxy-20α-dihydroprogesterone was not inhibitory. These investigators (64) proposed that decreased activity of 20α-hydroxysteroid dehydrogenase at puberty may permit the 17-hydroxylation of pregnenolone and progesterone, with a consequent formation of testosterone and other testicular androgens.

Other workers have provided data which supports this hypothesis. Berg et al. (62) reported a high ratio of 20α-dihydroprogesterone to 17-hydroxyprogesterone in prepubertal males from 5 to 7 years of age, with the ratio decreasing in older males from 11 to 15 years. Nayfeh et al. (60) reported a higher 20α-dihydroprogesterone to testosterone ratio in testicular tissue of prepubertal boys compared to normal adults.

To summarize, there are apparent differences in the ratios of 17-hydroxylated products to 20-hydroxylated products which change during the period of development in the human male. It is likely that there is a decrease in the activity of 20α-hydroxysteroid dehydrogenase at the time of puberty which permits the increased 17-hydroxylation of pregnenolone and progesterone and which thereby results in an increased synthesis and secretion of testicular androgens. These androgens may then be involved in some yet undetermined manner in the initiation of puberty.

Testicular Steroidogenesis in Subhuman Primates

In closing this discussion of testicular steroidogenesis in the human, it would seem appropriate to briefly outline the results obtained in similar studies with subhuman primates.

It has been demonstrated that there are substantial differences in the *in vitro* synthesis of androgens by the testis of a number of subhuman primates. Dorfman et al. (65) in an early study reported that in the Capuchin monkey, progesterone was metabolised primarily to 20α-dihydroprogesterone, with a much lower conversion to 17-hydroxyprogesterone and only a minor conversion to androstenedione and testosterone. The testis of the adult Capuchin monkey therefore is similar to the prepubertal human testis in that there is an elevated activity of 20α-hydroxysteroid dehydrogenase. 17-Hydroxyprogesterone was converted into 17-hydroxy-20α-dihydroprogesterone with a lower conversion into testosterone (65).

In the marmoset Saguinus oedipus, pregnenolone was converted into testosterone exclusively through the 4-ene pathway, with 17-hydroxyprogesterone the major metabolite (66). Progesterone, androstenedione and testosterone were also identified but no intermediates of the 5-ene pathway. However, it is interesting that in subsequent studies (67) 17-hydroxypregnenolone and dehydroepiandrosterone were more readily converted into testosterone and androstenedione than were progesterone and 17-hydroxyprogesterone. It appears that in this marmoset the lyase is either rate-limiting for the 4-ene pathway or the activity of the enzyme may be deficient as has been reported in Klinefelter's patients (45) and in the prepubertal human testis (61). The common marmoset Callithrix jacchus also formed testosterone from pregnenolone through the 4-ene pathway (68). However, in Callithrix, progesterone and not 17-hydroxyprogesterone was the major metabolite of pregnenolone, suggesting that 17-hydroxylase may be the rate-limiting step.

The testis of the adult baboon Papio anubis converted pregnenolone to testosterone through the 4-ene pathway, with only a minor conversion through the 5-ene pathway (69). Progesterone was by far the major metabolite of pregnenolone and with progesterone as substrate, most remained unmetabolised. The adult baboon testis also exhibited 20α-hydroxysteroid dehydrogenase and 5α-reductase activities. In the immature baboon, pregnenolone was also converted primarily into progesterone and then into testosterone, but the rate of conversion was substantially reduced (70). Also, there was an apparent lack of 5α-reductase activity since no 5α-reduced metabolites were identified. However, there apparently was 20α-hydroxysteroid dehydrogenase activity since 20α-dihydroprogesterone was identified as a metabolite of pregnenolone. The baboon therefore may be similar to the human regarding testicular 5α-reductase since there was little or no 5α-reductase activity in prepubertal individuals, but there did appear to be elevated 5α-reductase activity in adults.

Although previous workers have reported that in the Rhesus monkey Macaca mulatta there is either a 4-ene (71) or a 5-ene (72) pathway for testosterone synthesis, results from our laboratory suggested that both pathways are utilized by this species (73); progesterone was the predominant metabolite formed from pregnenolone, with 17-hydroxyprogesterone, androstenedione, testosterone, 17-hydroxypregnenolone and dehydroepiandrosterone as minor metabolites.Since progesterone accumulated in the pregnenolone incubates, and since progesterone when utilized as a substrate remained largely unconverted, 17-hydroxylase may in fact function as the rate-limiting enzyme for the synthesis of testosterone in the Rhesus monkey.

In the testis of the orang-utan Pongo pygaeus, pregnenolone is converted into testosterone through both pathways with the 5-ene pathway predominating (74). With pregnenolone as substrate, 17-hydroxypregnenolone and dehydroepiandrosterone were identified as primary metabolites, while progesterone and 17-hydroxyprogesterone were minor metabolites. Progesterone was recovered as 17-hydroxyprogesterone, androstenedione and testosterone. In contrast to other subhuman primates, the predominant metabolite formed from both pregnenolone and progesterone by the testis of the orang-utan was testosterone.

In summary, testicular steroidogenesis in subhuman primates may parallel the phylogenetic order. The more primitive primates such as the marmoset utilize an exclusive 4-ene pathway and this has also been demonstrated in several rodent species. As the phylogenetic order is ascended, subhuman primates such as the Rhesus monkey utilize both pathways and finally, the great apes, which are phylogenetically most similar to the human, may also utilize both pathways but with the 5-ene predominant. The higher subhuman primates may therefore serve as a suitable model for investigating the regulation of steroidogenesis in the human testis.

Concluding Remarks

The human testis is structurally organized to efficiently convert cholesterol to pregnenolone and then to testosterone through predominantly the 5-ene pathway. The 4-ene pathway may also be important. However, the 4-ene conversion may reflect testicular steroidogenic enzymes acting upon multiple steroid substrates. The predominant androgen synthesized and secreted by the normal human testis is testosterone. In a number of testicular disorders there appears to be suppression of the activities of specific steroidogenic enzymes which result in diminished testosterone production and an accumulation of specific intermediate metabolites. There are age-related differences in the activities of several steroidogenic enzymes which result in an alteration in the metabolites formed by the testis of individuals at different stages of life. Testicular enzymic activity may also be involved in the initiation of puberty in the human. The testis of higher subhuman primates may be utilized as a model to investigate the mechanism and regulation of testicular steroidogenesis in the human.

References

1. Sere, P.A., Chaikoff, I.L., Treitman, S.S., Burstein, L.S.: The extra hepatic synthesis of cholesterol. J. biol. Chem. 182, 629-634 (1950).
2. Brady, R.O.: Biosynthesis of radioactive testosterone *in vitro*. J. biol. Chem. 193, 145-148 (1951).
3. Kahnt, F.W., Neher, R., Schmid, K., Wettstein, A.: Bindung von 17-hydroxy-Δ^5pregnenolon und 3β-hydroxy-17-keto-Δ^5-androstein in nebennieren und testes-Gewebe. Experientia 17, 19-21 (1961).
4. Slaunwhite, W.R., Samuels, L.T.: Progesterone as a precursor of testicular androgens. J. biol. Chem. 220, 341-352 (1956).
5. Turner, C.D., Bagnara, J.T.: Endocrinology of the testis. In: "General Endocrinology", J.W.B. Saunders Co., Philadelphia, pp. 406-449 (1976).
6. Pelliniemi, L.J., Dym, M., Crigler, J.F., Retik, A.B., Fawcett, D.W.: Development of Leydig cells in human fetuses and in patients with androgen insensitivity. In: "Testicular Development, Structure and Function", Eds. Steinberger, A., Steinberger, E., Raven Press, New York, pp. 49-54 (1980).
7. Bakker, C.P., von der Plank-van Wisen, M.P.I., van der Molen, H.J.: Effect of cytosol fractions from lutropin-stimulated rat testes on pregnenolone production by mitochondria from normal rat testes. Biochim. biophys. Acta 543, 235-242 (1978).
8. Hall, P.F.: Endocrinology of the testis. In: "The Testis", Eds. Johnson, A.J., Gomes, E.R., Von Demark, N.L., Academic Press, New York, pp. 1-71 (1970).
9. Takatori, T., Privett, O.S.: Effect of hypophysectomy on cholesterol ester synthesis and hydrolysis in testes and on serum lecithin-cholesterol acetyl-transferase activity in rats. Endocrinology 103, 748-751 (1978).

10. Forchielli, E., Menon, K.M.J., Dorfman, R.I.: Control of testicular steroid biosynthesis. In: "The Gonads", Ed. McKerns, K.W., Appleton-Century-Crofts, New York, pp. 519-537 (1969).

11. Axelrod, L.R.: Metabolic patterns of steroid biosynthesis in young and aged human testes. Biochim. biophys. Acta 97, 551-556 (1965).

12. Yanaihara, T., Troen, P.: Studies of the human testes. I. Biosynthetic pathways for androgen formation in human testicular tissue *in vitro*. J. clin. Endocrinol. Metab. 34, 783-792 (1972).

13. Oshima, H., Fan, D.F., Troen, P.: Studies of the human testis. VI. NADH-linked reactions of microsomal steroid 20α- and 20β-hydroxysteroid dehydrogenase and 17α-hydroxylase. J. clin. Endocrinol. Metab. 40, 426-434 (1975).

14. Tamaoki, B.I.: General Review: Steroidogenesis and cell structure. Biochemical pursuit of sites of steroid biosynthesis. J. Steroid Biochem. 4, 89-118 (1973).

15. Fevold, A.R., Drummond, H.B.: Factors affecting the adrenocorticotropic hormone stimulation of rabbit adrenal 17α-hydroxylase activity. Biochim. biophys. Acta 313, 211-220 (1973).

16. Kremers, P.: Progesterone and pregnenolone 17α-hydroxylase: substrate specificity and selective inhibition of 17α-hydroxylated products. J. Steroid Biochem. 7, 571-575 (1976).

17. Lynn, W.S., Brown, R.H.: The conversion of progesterone to androgens by testes. J. biol. Chem. 232, 1015-1030 (1958).

18. Oshima, H., Nankin, H.R., Troen, P., Yoshida, K., Ochiai, K.: Leydig cell number and function in infertile men. In: "The Testis in Normal and Infertile Men", Eds. Troen, P., Nankin, H.R., Raven Press, New York, pp. 445-455 (1977).

19. Oshima, H., Ochiai, K.: On testicular 17β-hydroxysteroid oxidoreductase: Product activation of testosterone formation from androstenedione *in vitro*. Biochim. biophys. Acta 306, 227-236 (1973).

20. Oshima, H., Sarada, T., Ochiai, K., Tamaoki, B.I.: Intracellular distribution and properties of steroid 16α-hydroxylase in human testes. J. clin. Endocrinol. Metab. 27, 1249-1254 (1967).

21. Inano, H., Tamaoki, B.I.: *In vitro* effects of 16α-hydroxyprogesterone on the enzyme activities related to androgen production in human testes. Acta Endocrinol. 88, 767-777 (1978).

22. Eik-Nes, K.B.: Biosynthesis and secretion of testicular steroids. In: "Handbook of Physiology, Section 7, Endocrinology, volume V. Male Reproductive System", Eds. Hamilton, D.W., Greep, R.O., Amer. Physiol. Soc., Washington, pp. 95-115 (1975).

23. Faiman, C., Winter, J.S.D.: Diurnal cycles in plasma FSH, testosterone and cortisol in men. J. clin. Endocrinol. Metab. 33, 186-192 (1971).

24. Judd, H.L., Parker, D.C., Rakoff, J.S., Hopper, B.R., Yen, S.S.C.: Elucidation of mechanisms of the nocturnal rise of testosterone in men. J. clin. Endocrinol. Metab. 38, 134-141 (1973).

25. Hunter, J.: A treatise on the blood, inflammation and gun-shot wounds. Ann. R. Coll. Surg. England 54, 149-150 (1974).

26. Ruzicka, L.: Ueber die Synthese in Testikelhormones und Stereo-isomerer descelben durch abbau hydrierter Sterine. Helv. chim. Acta 17, 1395-1406 (1934).

27. Savard, K., Dorfman, R.I., Poutasse, E.: Biogenesis of androgens in the human testis. J. clin. Endocr. 12, 935 (1952).

28. Tamaoki, B., Inano, H., Nakano, H.: *In vitro* synthesis and conversion of androgens in testicular tissue. In: "The Gonads", Ed. McKerns, K.W., Appleton-Century-Crofts, New York, pp. 547-613 (1969).

29. Bell, J.B.G.: *In vitro* testosterone production from endogenous precursors by the seminiferous tubules and interstitium of the human testis. Steroid Lipid Res. 3, 315-320 (1972).

30. Hatakeyama, S.: A study on the interstitial cells of the human testis, especially on their fine-structural pathology. Acta path.Japon 15, 155-197 (1965).

31. Hatakeyama, S., Oshima, H.: Dual origin of Leydig cells in mouse testis with special reference to their cytogenetic correlation to macrophage. Gumma Symp. Endocrinol. 13, 81-93 (1976).

32. Van der Vusse, G.J., Kalkman, M.I., Van der Molen, H.J.: Endogenous production of steroid by subcellular fractions from total rat testis and from isolated interstitial tissues and seminiferous tubules. Biochim. biophys. Acta 297, 185-197 (1973).

33. Murota, S., Shikita, M., Tamaoki, B.: Androgen formation in the testicular tissues of patients with prostatic carcinoma. Biochim. biophys. Acta 117, 241-246 (1966).

34. Gloyna, R.E., Wilson, J.D.: A comparative study of the conversion of testosterone to 17β-hydroxy-5α-androstan-3-one by prostate and epididymis. J. clin. Endocr. 2, 970-977 (1969).

35. Bruchovsky, M., Wilson, J.D.: The conversion of testosterone to 5α-androstan-17β-ol-3-one by rat prostate *in vivo* and *in vitro*. J. biol. Chem. 243, 2012-2021 (1968).

36. Anderson, K.M., Liao, S.: Selective retention of dihydrotestosterone by prostatic nuclei. Nature 219, 277-279 (1968).

37. Murota, S., Shikita, M., Tamaoki, B.: Androgen formation in the testicular tissue of patients with prostatic carcinoma. Biochim. biophys. Acta 117, 241-246 (1966).

38. Oshima, H., Sarada, T., Ochiai, K., Tamaoki, B.: Effects of a synthetic estrogen upon steroid bioconversion *in vitro* in testes of patients with prostatic cancer. Invest. Urol. 12, 43-49 (1979).

39. Steinberger, E., Ficher, M., Smith, K.D.: Relation of *in vitro* metabolism of steroids in human testicular tissue to histologic and clinical findings. In: "The Human Testis", Eds. Rosemberg, E., Paulsen, C.A., Plenum Press, New York, pp. 439-458 (1970).

40. Bell, J.B.G., Lacy, D.: Studies on the structure and function of the mammalian testes. V. Steroid metabolism by isolated interstitium and seminiferous tubules of the human testes. Proc. roy. Soc. B. 186, 99-120 (1974).

41. Bell, J.B.G.: Studies of *in vitro* steroid metabolism by testis tissue from complete and incomplete forms of testicular feminization. Clin. Endocr. 4, 343-356 (1975).

42. Southren, A.L., Ross, R., Sharma, D.C., Gorden, G., Weingold, A.B., Dorfman, R.I.: Plasma concentration and biosynthesis of testosterone in the syndrome of feminizing testis. J. clin. Endocr. 25, 518-525 (1965).

43. Danezis, J.: Biosynthesis of androgens from progesterone by minute amounts of rat and human testicular tissue *in vitro*. Int. J. Fertil. 12, 308-311 (1967).

44. Schoen, E.J.: 17-Hydroxysteroid dehydrogenase activity in human testis. Acta Endocrinol. 56, 56-64 (1967).

45. Steinberger, E., Smith, K.D., Tcholakian, R.K., Chowdhury, M., Steinberger, A., Ficher, M., Paulsen, C.A.: Steroidogenesis in the human testis. In: "Male Infertility and Sterility", Eds. Mancini, R.E., Martini, L., Academic Press, New York, pp. 149-174 (1973).

46. Sulcova, J., Starka, L.: The metabolism of androgens in normal human testis and epididymis *in vitro*. Endocr. Exp. 7, 113 (1973).

47. Kjessler, B., Berg, A.A.: *In vitro* metabolism of ^{3}H-progesterone in human testicular tissue. Acta Endocrinol. Suppl. 207, 3-22 (1976).

48. Rosner, J.M., Macombe, J.C.: Biosynthesis of 5-androstenediol by human testis *in vitro*. Steroids 15, 181-193 (1970).

49. Rodriguez-Rigau, L.J., Weiss, D.B., Smith, K.D., Steinberger, E.: Suggestion of abnormal testicular steroidogenesis in some oligospermic men. Acta Endocrinol. 87, 400-412 (1978).

50. Sharma, D.C., Gabrilove, J.L.: Biosynthesis of testosterone and estrogens *in vitro* by the testicular tissue from patients with Klinefelter's Syndrome. Acta Endocrinol. 66, 737-744 (1971).

51. Lipsett, M.B., Davis, T.E., Wilson, H., Canfield, C.F.: Testosterone production in chromatin-positive Klinefelter's Syndrome. J. clin. Endocrinol. Metab. 25, 1027-1029 (1965).

52. Ahmad, K.N., Dykes, J.R.W., Ferguson-Smith, M.A., Lennox, B., Mack, W.S.: Leydig cell volume in chromatin-positive Klinefelter's Syndrome. J. clin. Endocrinol. Metab. 33, 517-520 (1971).

53. Schindler, A.E.: Steroid metabolism in the gonads of a patient with testicular feminization. Endokrinologie 65, 145-153 (1975).

54. Slaunwhite, W.R., Sandberg, A.A., Jackson, J.E., Staubitz, W.J.: Effects of estrogen and HCG on androgen synthesis by human testis. J. clin. Endocrinol. Metab. 22, 992-995 (1962).

55. Tamaoki, B.I., Shikita, M.: Biosynthesis of steroids in testicular tissue *in vitro*. In: "Steroid Dynamics", Eds. Pincus, G., Nakao, T., Tait, J.F., Academic Press, New York, pp. 493-530 (1966).

56. Rodriguez-Rigau, L.J., Tcholakian, R.K., Smith, K.D., Steinberger, E.: *In vitro* steroid metabolic studies in human testes. I: Effects of estrogen on progesterone metabolism. Steroids 29, 771-786 (1977).

57. Rodriguez-Rigau, L.J., Tcholakian, R.K., Smith, K.D., Steinberger, E.: *In vitro* steroid metabolic studies in human testes. II: Metabolism of cholesterol, pregnenolone, progesterone, androstenedione and testosterone by testes of an estrogen-treated man. Steroids 30, 729-739 (1977).

58. Steinberger, A., Ficher, M., Steinberger, E.: Studies of spermatogenesis and steroid metabolism in cultures of human testicular tissue. In: "The Human Testis", Eds. Rosemberg, E., Paulsen, C.A., Plenum Press, New York, pp. 333-343 (1970).

59. Kelch, R.P., Lindholm, U.B., Jaffe, R.B.: Testosterone metabolism of target tissues: 2. Human fetal and adult reproductive tissues, perineal skin and skeletal muscle. J. clin. Endocrinol. Metab. 32, 449-456 (1971).

60. Nayfeh, S.N., Coffey, J.C., Hansson, V., French, F.S.: Maturational changes in testicular steroidogenesis: hormonal regulation of 5α-reductase. J. Steroid Biochem. 6, 329-335 (1975).

61. Richards, G., Neville, A.M.: Steroid metabolism by the prepubertal human testis. Proc. Soc. Endocrinol. 61, xviii-xix (1974).

62. Berg, A.A., Kjessler, B., Lundkvist, K.: *In vitro* metabolism of ^{3}H-progesterone in human testicular tissue: II. Pubertal and adolescent boys. Acta Endocrinol. Suppl. 207, 23-35 (1976).

63. Sharma, D.C., Racz, E.A., Dorfman, R.I., Schoen, E.J.: A comparative study of the biosynthesis of testosterone by human testes and a virilizing interstitial cell tumor. Acta Endocrinol. 56, 726-736 (1976).

64. Fan, D.F., Oshima, H., Troen, B., Troen, P.: Studies on the human testis. IV. Testicular 20α-hydroxysteroid dehydrogenase and steroid 17α-hydroxylase. Biochim. biophys. Acta 360, 88-99 (1974).

65. Dorfman, R.I., Menon, K.M.J., Sharma, D.C., Joshi, S., Forchielli, E.: Steroid hormone biosynthesis in rat, rabbit and Capuchine testis. In: "Ciba Foundation Colloquia on Endocrinology of the Testis", Eds. Wolstenhome, G.E.W., O'Connor, M., Little, Brown and Co., Boston, volume 16, pp. 91-98 (1967).

66. Preslock, J.P., Steinberger, E.: Pathway of testosterone biosynthesis in the testis of the marmoset Saguinus oedipus. Steroids 28, 775-784 (1976).

67. Preslock, J.P., Steinberger, E.: Androgen biosynthesis by marmoset testis *in vitro*. Gen. Comp. Endocrinol. 31, 101-105 (1977).

68. Preslock, J.P., Steinberger, E.: Testicular steroidogenesis in the common marmoset Callithrix jacchus. Biol. Reprod. 17, 289-293 (1977).

69. Preslock, J.P., Steinberger, E.: Testicular steroidogenesis in the mature and immature baboon Papio anubis. Gen. Comp. Endocrinol. 33, 547-553 (1977).

70. Preslock, J.P., Steinberger, E.: Testicular steroidogenesis in the baboon Papio anubis. Steroids 32, 187-201 (1978).

71. Sharma, D.C., Joshi, S.G., Dorfman, R.I.: Biosynthesis of testosterone by monkey testes *in vitro*. Endocrinology 80, 499-504 (1967).

72. Hoschoian, J.C., Brownie, A.C.: Pathway for androgen synthesis in monkey testis. Steroids 10, 49-69 (1967).

73. Preslock, J.P., Steinberger, E.: Testicular steroidogenesis in the Rhesus monkey Macaca mulatta. Steroids 34, 527-535 (1979).

74. Preslock, J.P.: Testicular steroidogenesis in the orang-utan Pongo pygaeus. Proc. Soc. Study Reprod. 12th Annual Meeting, Abstr. 95 (1975).

HORMONE SECRETION AND STEROIDOGENESIS IN THE POLYCYSTIC OVARY SYNDROME

V. B. Mahesh
Department of Endocrinology, School of Medicine, Medical College of Georgia, Augusta, Georgia 30912, U.S.A.

The polycystic ovary syndrome is a somewhat vaguely defined clinical entity in women that is associated with excessive androgen secretion, hirsutism, obesity, oligomenorrhea or ovulatory failure and the presence of polycystic ovaries. In a small number of women there appears to be evidence of a genetic transmission of the disease. The hypersecretion of androgens by the adrenal or the ovary is well documented and the question of the presence of partial enzymatic abnormalities in adrenal and ovarian steroidogenesis has been raised repeatedly. This chapter discusses the evidence for and against the presence of enzymatic abnormalities in steroidogenesis in the polycystic ovary syndrome.

Variations in Clinical and Anatomical Findings in the Polycystic Ovary Syndrome

The presence of polycystic ovaries in the human was first described by Chareau in 1844 and partial resection of the ovaries for the management of polycystic ovarian disease was reported in Europe by 1897. In 1935 Stein and Leventhal (1), drew attention to the presence of large, pale polycystic ovaries in association with menstrual irregularities, ovulatory failure, infertility, hirsutism and obesity. Thus the association of the clinical symptoms and the presence of polycystic ovaries described a syndrome which was named after Stein and Leventhal.

Steroid Converting Enzymes and Diseases

In spite of extensive investigation of the polycystic ovary syndrome, a name commonly used because of the several variations in clinical findings from the description of Stein and Leventhal, its pathophysiology is still poorly understood. Many excellent reviews have appeared on the subject (2-16). Perhaps one reason for the lack of general agreement on the pathophysiology of the polycystic ovary syndrome is the fact that it is not a distinct entity.

The association of hirsutism, obesity, ovulatory failure and polycystic ovaries has long been recognized in infertile women. In a review published in 1962 of 1079 cases of surgically proven cases of polycystic ovarian disease, amenorrhea was found in 51% of the cases, hirsutism in 69%, obesity in 41%, infertility in 74% and virilization in 21% (17). Surprisingly 12% of the patients had cyclic menses and a corpus luteum was found at operation in 22%. Ovarian size and histology were also found to vary. In a study of 301 cases of polycystic ovarian disease, 40% of the patients had normal size ovaries and 46% did not have any thickening of the tunica (18). Of those patients with a thickened tunica, only 68 had the classical symptoms of hirsutism, obesity and menstrual irregularities or failure. Of these, 45 patients had enlarged ovaries while the remaining had normal size ovaries. It is obvious from this description that the polycystic ovary syndrome as described by clinical signs and symptoms is not a well-defined entity. Therefore a single concept of pathophysiology would not be applicable.

Polycystic Ovary Syndrome as a Genetic Disorder

There have been numerous reports in which more than one member of a family was affected by the polycystic ovary syndrome leading to the speculation of a genetic disorder. These reports include involvement of sisters (19-22), identical twins (17, 23, 24) and a mother and a daughter (25, 26). Vague et al. (27)

found a male sex chromatin pattern in the buccal smear of one of four patients with the Stein-Leventhal Syndrome. The findings of chromosomal abnormalities in the polycystic ovary syndrome were confirmed by Wais (28) and Netter et al. (29). Abnormal ovarian steroid secretion by polycystic ovaries was therefore considered to have a genetic origin (30). Normal sex chromatin and karyotypes were found by Byrd et al. (31) in 9 patients with the Stein-Leventhal Syndrome in whom the presence of large pale polycystic ovaries was confirmed by laparotomy. In a study of 18 families with Stein-Leventhal Syndrome, Cooper et al. (32) concluded that although a typical and consistent chromosomal abnormality is not associated with the syndrome, there appeared to be an autosomal dominant mode of transmission. The genetic basis of some cases of polycystic ovarian disease possibly linked with the X-chromosome has been reported (33-35). In one of these studies hirsutism was found to occur in five generations with bilaterally enlarged polycystic ovaries in three women in different sibships of two generations (34). A brother of one of the women with polycystic ovaries had Klinefelter's Syndrome. Parker et al. (36) found a trisomy of chromosome 14 in 2 to 4% of the cells of 5 out of 15 patients with polycystic ovarian disease. In spite of this evidence, the genetic basis of polycystic ovarian disease is limited to only a small number of cases and is not a generalized finding.

The Adrenal and/or the Ovary as a Source of Excessive Androgens

The elevation of urinary 17-ketosteroids, pregnanetriol and Δ^5-pregnenetriol is well documented in the polycystic ovary syndrome and until the early 1960's the adrenal was considered to be the only source of these steroids (37-41). This led to the speculation that altered adrenal steroid secretion patterns were the cause of polycystic ovarian disease. This belief was strengthened by the widely reported occurrence of polycystic ovaries in association with adrenal tumors and with cases of congenital adrenal hyperplasia (42-50).

In 1953, Greenblatt (51) observed a fall in urinary 17-ketosteroids after wedge-resection in untreated and cortisone treated patients with the polycystic ovary syndrome and concluded that the ovary was a source of excessive androgens. These findings were subsequently confirmed by other investigators (52, 53). In an attempt to study the role of the ovary in androgen secretion, a detailed investigation was undertaken by Mahesh and coworkers (2, 4, 5, 6, 8, 9, 54, 55, 56, 57). This study included: measurement of individual urinary steroids before and after wedge-resection of the ovary and after adrenal suppression and ovarian stimulation, measurement of plasma androsterone sulfate and dehydroepiandrosterone sulfate before and after wedge-resection of the ovary, measurement of the steroid content of normal and polycystic ovarian tissue and the measurement of steroids in ovarian venous blood. The measurement of urinary 11-deoxy-17-ketosteroids, 11-oxygenated-17-ketosteroids and tetrahydrocorticoids in patients with the Stein-Leventhal Syndrome enabled them to be classified into three groups (Fig. 1) (56). Group 1 had an elevation of both urinary 11-deoxy-17-ketosteroids and 11-oxygenated-17-ketosteroids. Since 11-oxygenated-17-ketosteroids arise exclusively from the metabolism of steroids of adrenal origin, these patients were considered to have an adrenal disorder. Suppression of adrenal function by dexamethasone was effective in suppressing the 11-deoxy- and 11-oxygenated-17-ketosteroids. In Group 2 the only abnormality was the elevation of the 11-deoxy-17-ketosteroids (Fig. 2) which could arise from the adrenal and from the ovary. On adrenal suppression, Group 2 could be subdivided into Groups 2A and 2B. In Group 2A, dexamethasone administration lowered the levels of 11-deoxy-17-ketosteroids to those comparable to normal women treated with the adrenal suppressant and therefore the source of excessive androgens appeared to be the adrenal. In Group 2B dexamethasone administration lowered the 11-deoxy-17-ketosteroids only slightly and the residual 11-deoxy-17-ketosteroids were significantly higher

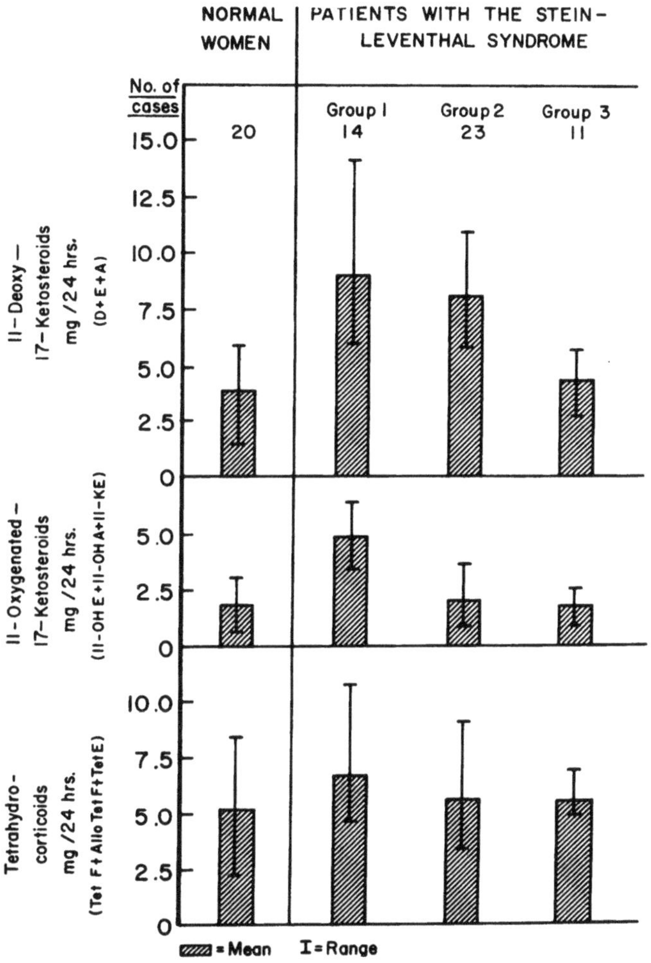

Fig. 1 Urinary 11-deoxy-17-ketosteroids, 11-oxygenated-17-ketosteroids and tetrahydrocorticoids in normal women and women with the polycystic ovary syndrome. The hirsute women are divided into three groups. Group 1 had elevated 11-deoxy and 11-oxygenated-17-ketosteroids, Group 2 had only elevated 11-deoxy-17-ketosteroids and Group 3 had normal levels of all steroids. (From 56)

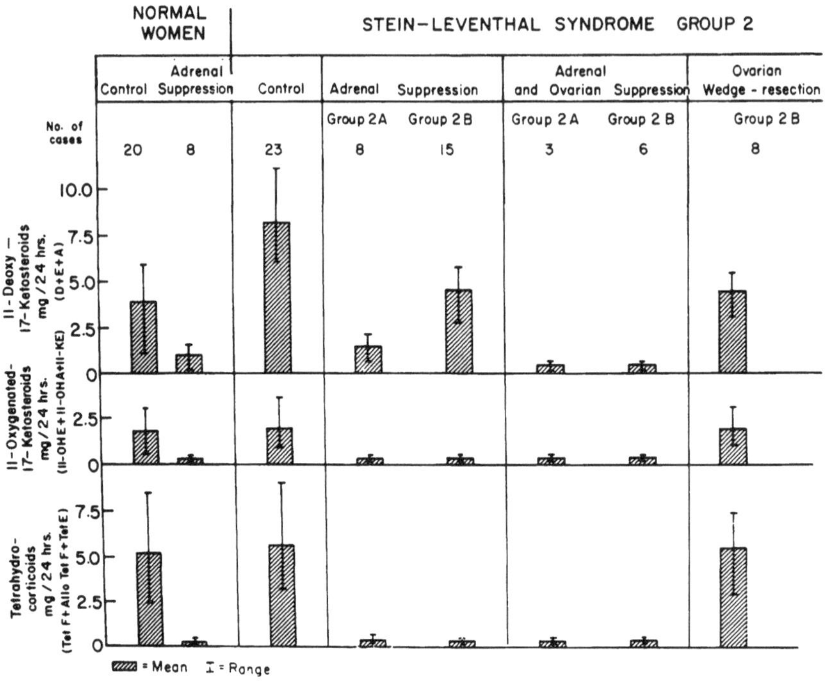

Fig. 2 Urinary steroids in normal women and patients with the polycystic ovary syndrome in Group 2 before and after adrenal and ovarian suppression and after wedge-resection of polycystic ovaries. (From 56).

than those in normal women suppressed with dexamethasone. Further suppression of these could be achieved by ovarian suppression with estrogens. These observations indicated that the ovaries were responsible in Group 2B for excessive androgen secretion. Further confirmation of an ovarian source of androgens was obtained by a decrease in plasma androsterone sulfate and dehydroepiandrosterone sulfate after wedge-resection of the ovary (4) and a lowering of urinary 11-deoxy-17-ketosteroids after wedge-resection (Fig. 2). Patients with the polycystic ovary syndrome in Group 3 did not show any abnormality in steroid secretion as judged by the basal level of steroids as well as the effect of adrenal suppression.

Direct confirmation of an ovarian source of androgens in the polycystic ovary syndrome was obtained by isolation of large quantities of Δ^4-androstenedione and dehydroepiandrosterone from ovarian tissue and ovarian vein blood (Fig. 3) (2, 4, 55). The isolation of large quantities of dehydroepiandrosterone from polycystic ovaries was confirmed (58, 59) and large quantities of Δ^4-androstenedione were isolated from follicular fluid of polycystic ovaries (60, 61).

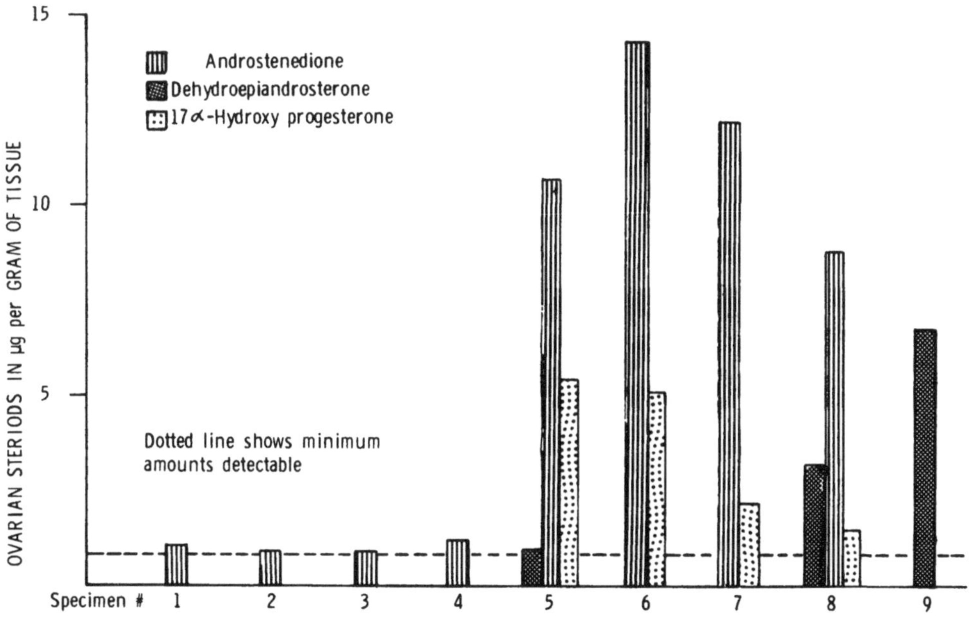

Fig. 3 Androstenedione, dehydroepiandrosterone and 17α-hydroxyprogesterone content of normal (Specimens 1-4) and polycystic ovaries (Specimens 5-9). (From 4).

Source of Excessive Androgens

It is now very well documented on the basis of urinary and blood steroids that there is excessive androgen secretion in hirsutism and the polycystic ovary syndrome. A review of the extensive work is beyond the scope of this chapter. The spectrum of steroids that are elevated include progesterone, 17α-hydroxyprogesterone, dehydroepiandrosterone, androstenedione, testosterone and 5α-dihydrotestosterone (Fig. 4) (62). Determination of the source of excessive androgens, however, is a difficult problem. The most common approach has been to measure the blood and/or urinary steroids before and after adrenal suppression (2, 4, 5, 8, 56, 63-73). However, direct effects of glucocorticoids on gonadotropin secretion have been demonstrated (74) in the human and the baboon. Furthermore, several investigators have concluded that adrenal suppression lowers ovarian androgen production (68, 69, 75-79). On the other hand, it was also shown that doses of 2 mg dexamethasone/day had no effect on ovarian steroid secretion (81, 82).

Combined adrenal and ovarian venography for determining the site of excessive androgen production was first used by Kirschner and Jacobs in 1971 (75) and subsequently by Stahl et al. (82). A detailed study to evaluate the usefulness of selective catheterization of adrenal and ovarian veins for determining the source of excessive androgens was subsequently carried out by Parker et al. (62). The purpose of this study was to determine whether sampling of ovarian and adrenal venous blood over short periods of time would accurately reflect steroidogenesis in view of the pulsatile and episodic nature of steroid secretion. Furthermore, it would be of interest to determine if venous blood from the left adrenal and left ovary would be representative of the contralateral adrenal or ovary. Finally, because hCG and ACTH has been used frequently for various stimulatory studies, any possible effect of ACTH on the ovary and hCG on the adrenal needed to be determined.

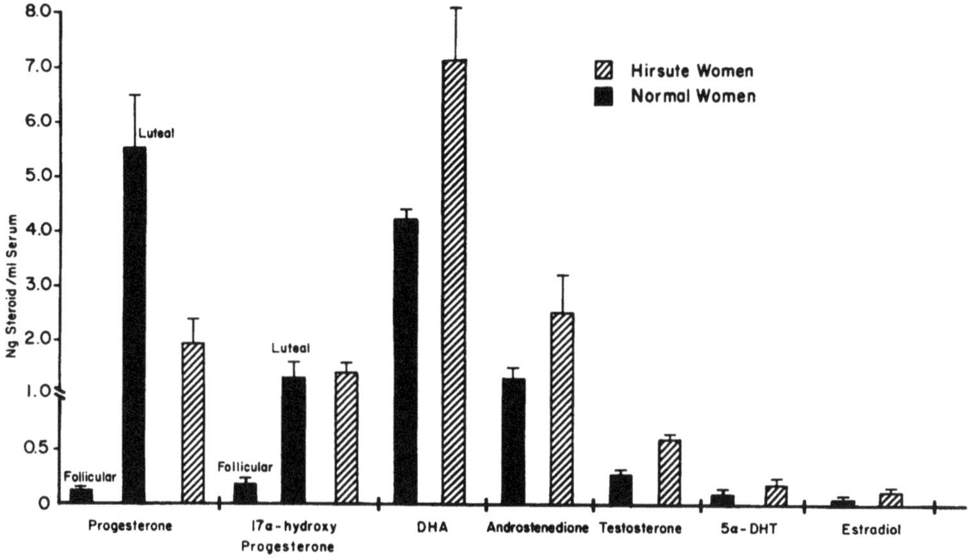

Fig. 4 Progesterone, 17α-hydroxyprogesterone, dehydroepiandrosterone, androstenedione, testosterone, 5α-dihydrotestosterone and estradiol levels in serum of normal and hirsute women. (From 62).

Blood samples from a peripheral vein, the left adrenal vein and the left ovarian vein immediately before and 10, 20 and 30 minutes after the injection of saline to a hirsute woman showed considerable pulsatile fluctuation in the steroid values at different time points (62). When 40 IU of ACTH were administered intravenously to a hirsute woman on the 7th day of her cycle, remarkable increases occurred in the progesterone, testosterone, Δ^4-androstenedione, 17α-hydroxyprogesterone and cortisol concentrations in the adrenal vein blood within 20 minutes (Figs. 5 and 6). Ovarian vein blood showed only small increases in progesterone (Fig. 6), dehydroepiandrosterone and Δ^4-androstenedione. There was no change in ovarian estradiol

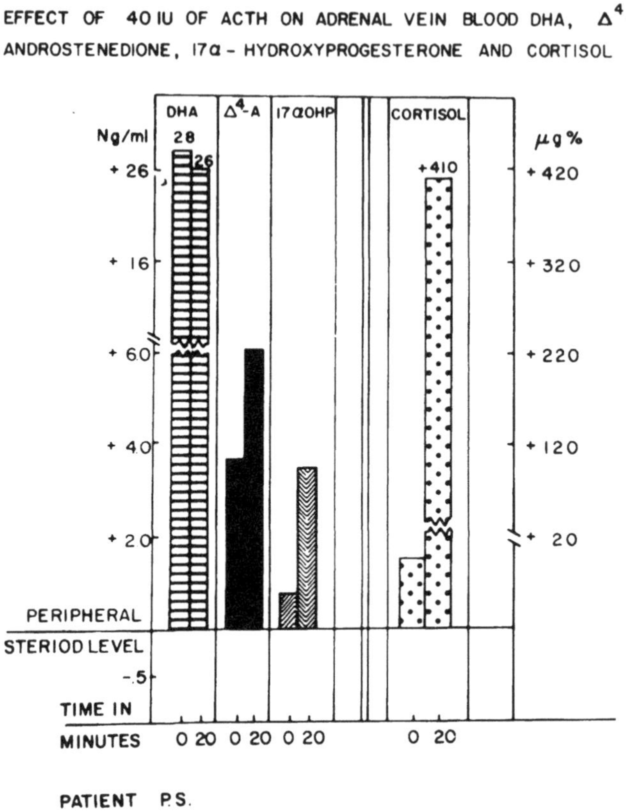

Fig. 5 Changes in Δ^4-androstenedione, 17α-hydroxy-progesterone and cortisol in adrenal vein blood 20 minutes after 40 IU ACTH with no change in free dehydroepiandrosterone. (From 62).

and testosterone as a result of ACTH administration. The increases in ovarian progesterone, dehydroepiandrosterone and Δ^4-androstenedione were in all probability due to the ovarian

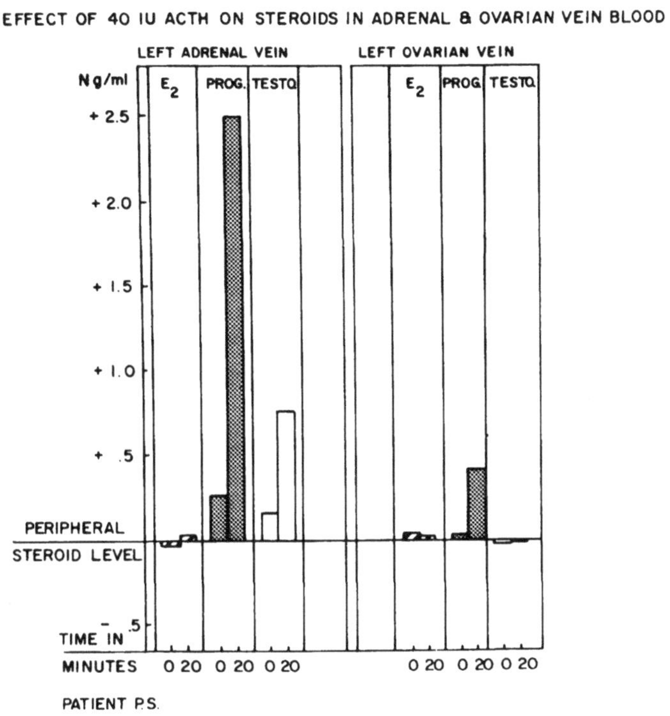

Fig. 6 ACTH increased progesterone and testosterone in adrenal vein blood with no changes in estradiol. Ovarian vein blood steroids unaffected by ACTH with the exception of a small change in progesterone. (From 62).

utilization of adrenal precursors rather than a direct effect of ACTH on the ovary.

In a hirsute patient with ovulatory failure and elevation of blood dehydroepiandrosterone, Δ^4-androstenedione and testosterone levels, 2000 IU hCG was administered and blood was collected from a peripheral vein, left ovarian vein and the left adrenal vein immediately before and 10, 20 and 30 minutes after

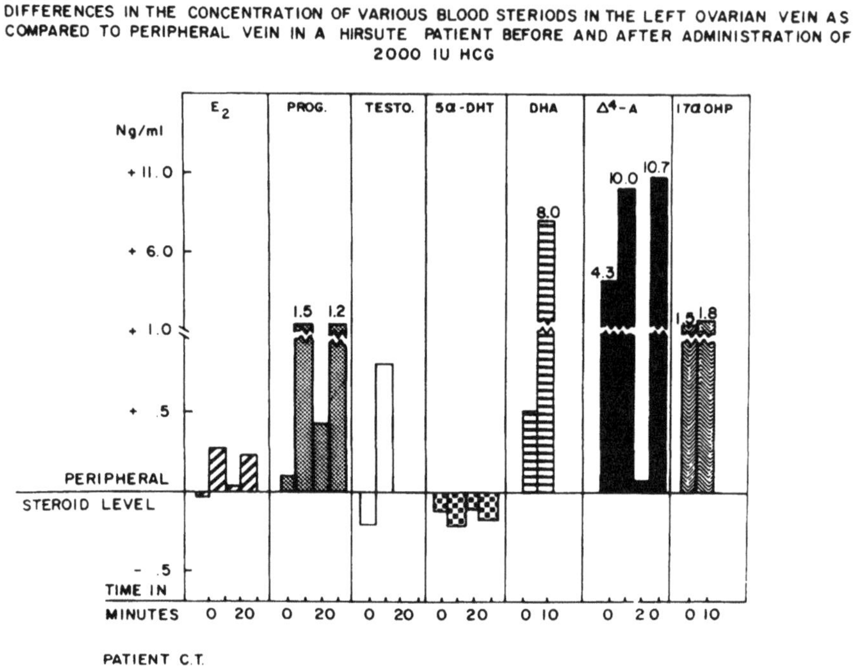

Fig. 7 Increases in estradiol, progesterone and Δ^4-androstenedione in left ovarian vein blood of a hirsute patient 10 and 30 minutes after 2,000 IU hCG. Testosterone and dehydroepiandrosterone also showed a marked rise at 10 minutes. A fall in the levels of estradiol, progesterone and Δ^4-androstenedione at 20 minutes emphasizes the pulsatile nature of steroid secretion in ovarian vein blood. (From 62).

hCG. The results (Fig. 7) show a rise in estradiol, progesterone, testosterone, dehydroepiandrosterone and Δ^4-androstenedione in the left ovarian vein within 10 minutes. A fall in estradiol, progesterone and Δ^4-androstenedione was observed at 20 minutes followed by another rise at 30 minutes. Minimal changes took place in the adrenal vein blood after hCG. These results emphasize the limitations of a single blood sample obtained from adrenal and ovarian veins due to the episodic nature of steroid secretion.

Ovarian and adrenal suppression with prednisone and estradiol was very effective in suppressing ovarian output of steroids and ovarian response to 2000 IU hCG (Fig. 8). In order

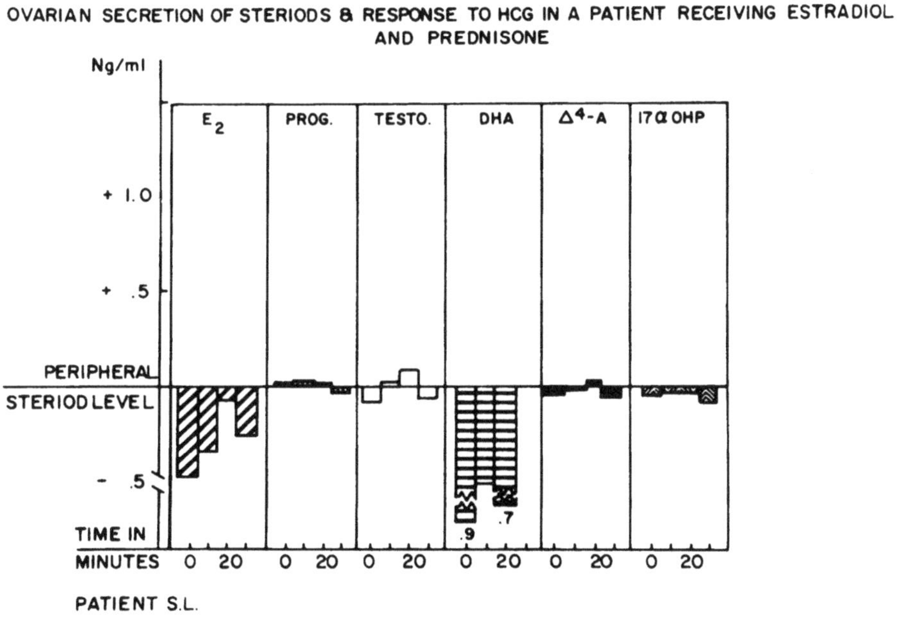

Fig. 8 Adrenal and ovarian suppression leading to low ovarian steroid secretion, and no response to hCG stimulation. (From 62).

to determine whether left ovarian vein blood and the responsiveness to hCG would be representative of both the ovaries, hCG was administered to a hirsute woman with a corpus luteum on the right ovary documented by laparoscopy. Ovarian vein blood was obtained from the left ovary immediately before and 10, 20 and 30 minutes after hCG. Prior to hCG stimulation, the left ovary secreted only dehydroepiandrosterone, testosterone, Δ^4-androstenedione and 17α-hydroxyprogesterone but not estradiol and progesterone (Fig. 9). As a result of hCG stimulation of the dominant right ovary, the peripheral concentration of the major ovarian steroids increased and the difference between peripheral steroids and left ovarian vein steroids showed remarkable decreases in ovarian output of steroids by the left ovary. It becomes clear from these results that the study of ovarian vein steroids from one ovary are not a reliable index of ovarian secretion if a corpus luteum is present on the contralateral ovary.

Kirschner (83) summarized his experience with adrenal and ovarian venous catheterization. Attempts to obtain left adrenal and left ovarian vein blood samples were successful in 88% and 78% of the cases attempted. It was considerably more difficult to obtain right adrenal and right ovarian vein blood samples, the success rate being 60% and 55% respectively. Blood from both adrenal veins was obtained successfully in 58% of the cases and both ovarian veins in 50% of the cases. Success rate for all four blood samples was only 40%. In view of the pulsatile nature of steroid secretion, the problem of stress and the comparatively low success rate in obtaining blood from all four adrenal and ovarian veins, this procedure cannot be recommended for routine use. Therefore the accurate evaluation of the adrenal and ovarian contribution to excessive androgen secretion still remains controversial.

Abnormal or Disturbed Hypothalamic-Pituitary Feedback System

Early reports on the measurement of urinary gonadotrophins by bioassay in patients with the polycystic ovary syndrome indicated a hypersecretion of LH (84, 85). This was confirmed by radioimmunoassays of serum LH, the levels of which were either

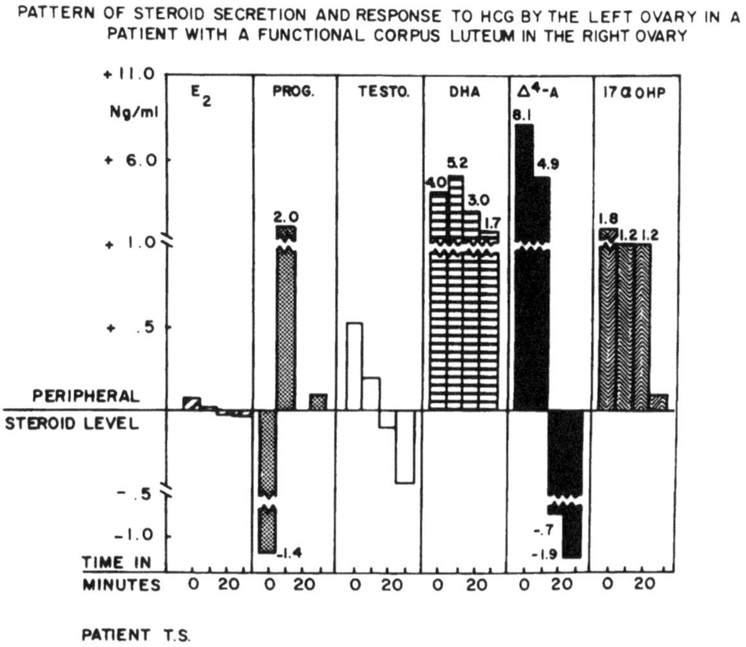

Fig. 9 When a patient with a corpus luteum on the right ovary was stimulated with 2,000 IU hCG, a decrease in the difference between peripheral blood and left ovarian venous blood steroids occurred in testosterone, dehydroepiandrosterone, Δ^4-androstenedione and 17α-hydroxyprogesterone. This was primarily due to the elevation of these steroids in peripheral blood because the right, but not the left, ovary responded to hCG. (From 62).

normal or elevated while the FSH levels were normal or depressed (86-93). The high LH concentrations found in patients with the polycystic ovary appear to be the result of increased frequency and/or a greater amplitude of pulsatile LH release (13, 93). Furthermore, when stimulated with LHRH, patients with the polycystic ovary syndrome showed an exaggerated response in the release of LH (13, 90-97). These observations have contributed to the concept of an abnormal hypothalamic-pituitary axis in the polycystic ovary syndrome. Although the pulsatile LH levels are elevated in the majority of cases with the polycystic ovary syndrome, normal levels of LH have invariably been found. Some investigators have attributed this to be due to infrequent sampling. Others have divided the patients with polycystic ovaries into typical or type I with elevated LH levels and enlarged ovaries and atypical or type II with normal LH and normal sized ovaries (90, 94, 98). However, many investigators have found no relationship between serum LH levels and ovarian size (99, 100).

An abnormal hypothalamic-pituitary axis in the polycystic ovary syndrome should result in a modification of the positive and negative feedback effect of estradiol on the release of gonadotrophins. This was not the case because the immediate lowering of LH after estradiol administration appeared to be comparable to that found during the normal cycle (13, 87, 93). Furthermore, the administration of 220 µg ethinyl estradiol gave a positive LH surge similar to that observed in normal women although it occurred a day earlier (93). Treatment of patients with clomiphene citrate usually results in a small increase of gonadotrophins during administration of the drug followed by a peak in estrogens that triggers the ovulatory surge of gonadotrophins (Fig. 10) (4, 88, 93, 101). This sequence of events is similar to that of the normal cycle. These observations do not support the hypothesis of a defective hypothalamic-pituitary axis in the polycystic ovary syndrome.

In view of the above, an explanation for the high pulsatile levels of LH and hyperresponsiveness to LHRH was necessary to rule out the concept of an abnormal hypothalamic-pituitary axis. It is now recognized that the responsiveness of the pituitary to LHRH undergoes extensive changes during the menstrual cycle which is at least in part related to circulating steroids in blood (102-105). Augmentation of pituitary responsiveness to

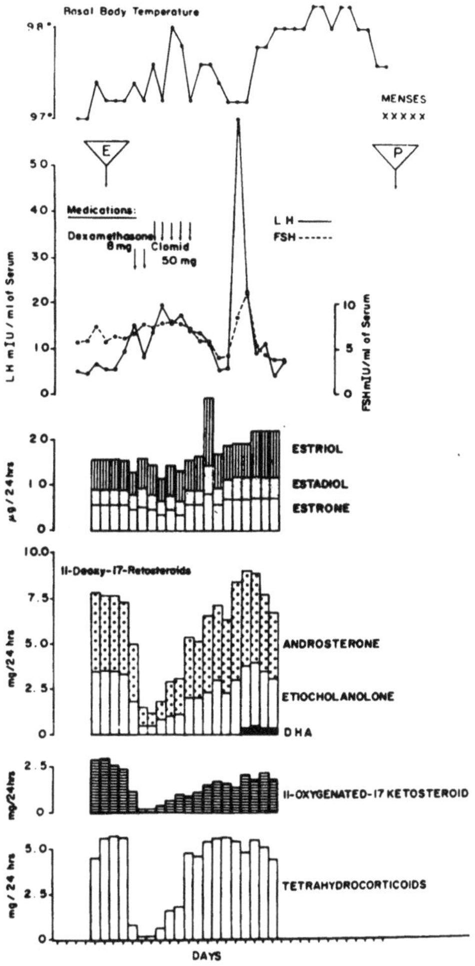

Fig. 10 Ovulation induced by clomiphene therapy in a 23-year-old married nulligravid woman who was anovulatory and gave a history of increasing hirsutism of five years duration. Note initial increase in LH during clomiphene therapy which was followed by an estrogen peak leading to the ovulatory surge of FSH and LH. (From 88).

LHRH by estrogens in the human is also well documented (102, 106, 107). Therefore, it is possible that in the polycystic ovary syndrome, the elevated LH and enhanced responsiveness to LHRH is caused by the elevated levels of circulating steroids in blood. Duignan et al. (92) reported that the LH correlated inversely with circulating testosterone in polycystic ovary syndrome while Mortimer et al. (108) reported a positive correlation of LH with testosterone. The correlation of circulating LH with estrogens is more widely recognized and accepted (13, 91, 93, 109, 110). The fractional increase of LH after LHRH is also related to the level of circulating estrogens (13). The administration of estrone benzoate to patients with the polycystic ovary syndrome further suppressed FSH without altering LH and thus enhanced the disparity between the two gonadotrophins (111).

An Animal Model that Demonstrates Inappropriate Gonadotrophin Secretion, Ovulatory Failure and Polycystic Ovaries

Various concepts of the pathogenesis of the polycystic ovary syndrome in the human involve an abnormal adrenal, an abnormal ovary and an abnormal hypothalamic-pituitary axis. The relationship between excessive androgen secretion and ovulatory failure with formation of polycystic ovaries has been observed in a variety of clinical conditions such as polycystic ovary syndrome, Cushing's Syndrome, congenital adrenal hyperplasia and arrhenoblastoma. In a large number of cases, reduction of circulating androgens by measures such as wedge-resection of the ovaries, glucocorticoids to suppress adrenal function or removal of the tumor has led to restoration of cyclic ovulatory function. Therefore, a detailed study of the effect of circulating androgens on ovulation was undertaken.

In order to study the effects of circulating androgens on ovulation in an animal with normal pituitary, hypothalamic, adrenal and ovarian function, the immature rat was chosen as a model because the ovary would be free of degenerating corpora lutea that could be influenced by the changing hormonal environment. The weak androgen dehydroepiandrosterone (DHA) was chosen for initial study because it was easier to choose a dose level that could be used in a group of prepubertal animals beginning to show variability in the exact time of onset of puberty. The administration of 6 mg DHA/100 g body wt. to 27-day-old immature rats resulted in precocious ovulation followed by ovulatory failure and the formation of polycystic ovaries (112, 113, 114). Blood gonadotrophin measurements during ovulatory failure showed low levels of LH, slightly elevated FSH and markedly elevated prolactin. The sequence of events in the DHA-induced precocious ovulation were an increase in circulating estradiol followed by a depletion of the pituitary and hypothalamic cytoplasmic estradiol receptors leading to the gonadotrophin surge responsible for ovulation (115). The events were similar to those observed in the natural onset of puberty (116) or the first ovulation induced by pregnant mare serum gonadotrophin (PMSG) administration to immature rats (117). Continued DHA treatment resulted in inappropriate gonadotrophin secretion and lack of replenishment of hypothalamic and pituitary cytoplasmic estradiol receptors (11). The withdrawal of DHA administration resulted in a gradual decline in blood steroids and a replenishment of pituitary and hypothalamic estradiol receptors which correlated well with the return of FSH and LH to levels found in the intact rat. This was followed by a return of cyclic ovulation and fertility. The period required for the return of cyclicity after DHA withdrawal was 7 to 19 days with a mean interval of 10 days.

The experiments with the immature rats raised two questions. Firstly, was the dose of DHA unphysiologically large and, secondly, would ovulatory failure and polycystic ovary formation occur in the adult rat. In order to answer these questions, the ability of various doses of DHA to prevent the post-castration rise of gonadotrophins in the adult rat was studied (118). The lowest dose that significantly lowered FSH and LH, even beyond that found in diestrus controls, was 3 mg/100 g body wt. The 1.5 mg/100 g body wt. dose was only able to significantly lower LH and FSH but the values were still considerably greater than the values in the diestrus controls. Various doses of DHA were then administered to cycling female rats for 20 days. It was found that a dose as low as 0.187 mg/100 g body wt. could disrupt normal cycles and lead to the formation of polycystic ovaries.

The above series of experiments demonstrated in animals with normal adrenals, ovaries and hypothalamic-pituitary function, that administration of a weak androgen could cause inappropriate gonadotrophin secretion and ovulatory failure. Although it is difficult to compare the human with a rat model due to species differences, the relationship between circulating androgens and ovulatory failure established in this rat model are of considerable interest.

Adrenal Enzymatic Abnormalities Leading to Inappropriate Gonadotrophin Secretion and Polycystic Ovaries

In view of the ability of the adrenal to secrete androgens and the association of polycystic ovaries with virilizing adrenal tumors and congenital adrenal hyperplasia (42-50), the adrenal has been frequently considered the source of excessive androgens. Therefore it is not surprising that a search for abnormal adrenal 21-hydroxylase, 11β-hydroxylase and 3β-hydroxysteroid dehydrogenase activity has been made in cases of hirsutism and polycystic ovaries. Since most patients with hirsutism and the polycystic ovary syndrome do not have the frank virilization found in congenital adrenal hyperplasia, an enzymatic abnormality that is attenuated was the object of the search. Givens et al. (73) observed that blood androstenedione underwent diurnal changes in hirsute women that correlated with the cortisol rhythm and androstenedione increased significantly on stimulation with 0.5 U ACTH. This small dose of ACTH caused an elevated ratio of blood 17α-hydroxyprogesterone to cortisol. They concluded that this indicated a mild degree of 21-hydroxylase or 11β-hydroxylase deficiency in the adrenal. An increase in androstenedione in the presence of a proportionally lower 17α-hydroxyprogesterone and cortisol led to the conclusion of the presence of a decreased level of C_{21}-3β-hydroxysteroid dehydrogenase. ACTH stimulation and the ratios of various blood steroids has been used by other investigators to demonstrate deficiencies of adrenal 11β-hydroxylase, 21-hydroxylase and 3β-hydroxysteroid dehydrogenase in hirsute women and women with the polycystic ovary syndrome. In several cases, an adrenal enzymatic abnormality has been associated with polycystic ovarian disease (119-124). However, Lachelin et al. (125) concluded that demonstration of adrenal enzymatic abnormalities in polycystic ovarian syndrome only occurred after prolonged rather than acute exposure to ACTH. Furthermore, it is not clear what triggers an adrenal enzymatic abnormality, if present, in the polycystic ovary syndrome; androgen inhibition of adrenal enzymes may be one mechanism (126, 127).

Ovarian Enzyme Defects in the Polycystic Ovary Syndrome

The question of the presence of enzymatic abnormalities in ovarian steroidogenesis has also been raised repeatedly. Measurement of androgens and estrogens in ovarian venous blood and a comparison of the estrogen/androgen ratio in venous blood from normal ovaries and polycystic ovaries clearly established the presence of a functional abnormality in ovarian steroidogenesis in polycystic ovaries (2, 4). Further evidence of an ovarian enzymatic abnormality was obtained by *in vitro* incubation studies with [^{3}H]-dehydroepiandrosterone and [4-^{14}C]-androstenedione (Fig. 11). Samples 8 and 9 which contained large quantities of dehydroepiandrosterone (Fig. 3) showed poor conversion of [^{3}H]-dehydroepiandrosterone to androstenedione and estrogens compared to a normal ovarian sample,

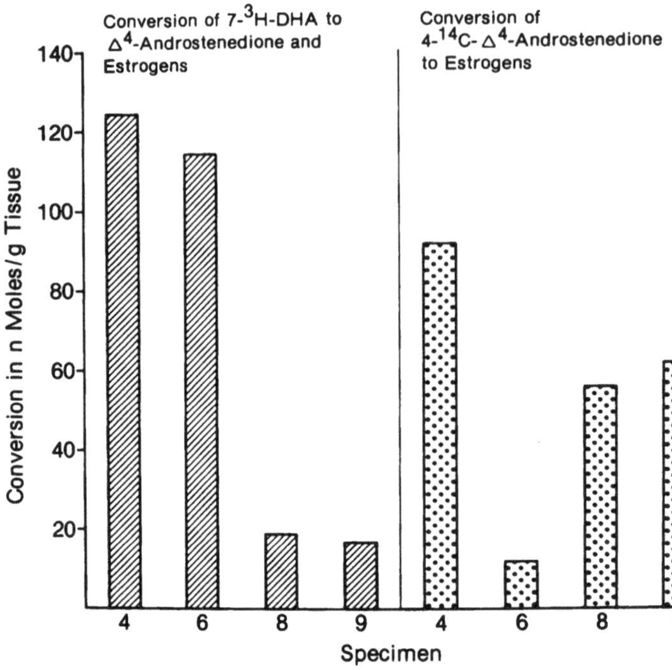

Fig. 11 Conversion of 7-^{3}H-dehydroepiandrosterone to Δ^{4}-androstenedione and estrogens and 4-^{14}C-Δ^{4}-androstenedione to estrogens by a normal ovarian tissue *in vitro* and three polycystic ovarian tissue specimens. Their steroid content is shown in Fig. 3. Note poor 3β-hydroxysteroid dehydrogenase activity in specimens 8 and 9 and poor aromatase in specimen 6. (From 2).

suggesting an abnormality of 3β-hydroxysteroid dehydrogenase (Fig. 11). Sample 6 on the other hand showed poor aromatization of androstenedione to estrogens. These enzymatic abnormalities in ovarian steroidogenesis are summarized in Fig. 12. Similar observations were made by Axelrod and Goldzieher with regard to *in vitro* ovarian steroidogenesis in human polycystic ovaries (128).

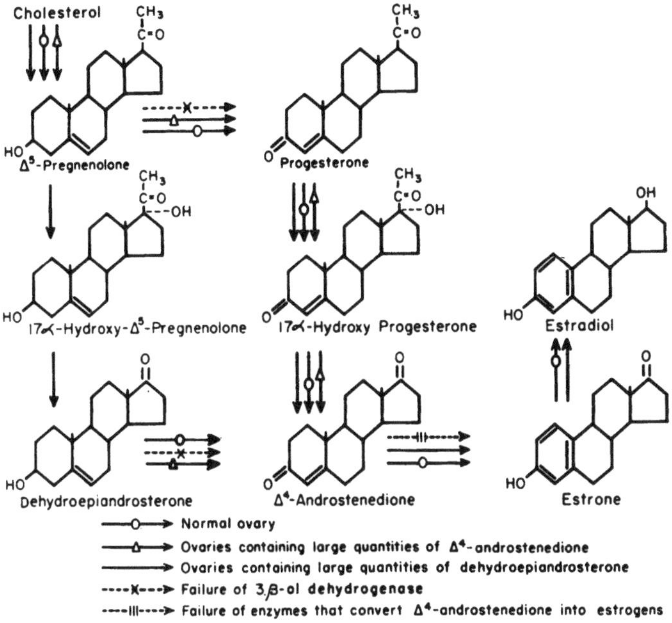

Fig. 12 Major pathways of biosynthesis of steroids in the ovary. (From 2).

Enzymatic abnormalities were also found using histochemical techniques (129-132).

In spite of the above evidence, it was not clear whether there was a basic defect in ovarian steroidogenesis in the polycystic ovary syndrome or only a functional defect due to inappropriate gonadotrophin secretion, acyclicity and lack of proper follicular development. Axelrod and Goldzieher (133) have reported a case of concurrent adrenal and ovarian 3β-hydroxysteroid dehydrogenase deficiency in a case of polycystic ovarian disease. The greater conversion of progesterone to androstenedione persisted in a patient with polycystic ovary even though the patient had responded to wedge-resection and was ovulating normally at the time of the second study (134).

A poor conversion of Δ^5-pregnenolone to Δ^4-3-ketosteroids was found in the ovaries of a patient with chronic familial anovulation who had earlier responded to wedge-resection for 4 months (135). An 11β-hydroxylase specific for 21-deoxysteroids has also been reported in polycystic ovarian tissue (136, 137). On the other hand, *in vitro* studies with granulosa cells and in some cases theca cells from normal and polycystic ovarian tissue has shown no difference in steroidogenesis (138, 139). Based on the small number of reports of a persistent alteration in ovarian steroidogenesis and *in vitro* studies, it appears unlikely that a basic rather than a functional ovarian enzyme defect is a frequent finding in polycystic ovarian disease.

Direct Androgen Effects on the Ovary

Androgen administration to an animal with normal hypothalamic-pituitary, adrenal and ovarian function can result in inappropriate gonadotrophin secretion, formation of polycystic ovaries and ovulatory failure (112-118). In the human, such an effect has been documented in cases of androgen excess from the adrenal due to enzymatic abnormalities of the adrenal (120-124).

Direct androgen effects on the ovarian follicle leading to follicular atresia have been demonstrated by several investigators (140-144). An androgen receptor in the granulosa cells of estrogen-treated hypophysectomized rats has also been identified (145, 146). It is therefore possible that excessive androgens may have a significant effect on follicular maturation and atresia in the polycystic ovary syndrome. In the human, wedge-resection of the ovary in the polycystic ovary syndrome did not bring about significant alterations in the circulating levels of FSH and LH until the first ovulation occurred (22, 147). In the absence of any significant change in circulating gonadotrophins, the ovary nonetheless was able to produce the estrogen trigger leading to the surge of gonadotrophins (22). Thus it was postulated that reduction of the intra-ovarian effect of androgens by wedge-resection contributed to the normal pattern of ovarian steroidogenesis.

Induction of ovulatory failure and formation of polycystic ovaries in rats by administration of dehydroepiandrosterone (DHA) has already been mentioned (112-118). It was of interest to us to determine if the ovaries of the DHA treated rat were active in steroidogenesis. Serum FSH and LH levels were measured in adult rats treated with 1.5 and 3 mg DHA/100 g body wt. for 15 or 20 days (14). The absence of change in the two groups could be due to stored DHA at the injection site. In other animals DHA was administered for 15 days followed by ovariectomy. A group of the ovariectomized rats received vehicle for 5 days while another group received DHA. A dramatic elevation of FSH and LH after ovariectomy was observed and even continued treatment with DHA did not restore the levels to those found in intact rats during DHA treatment (Fig. 13). These results indicated that the ovaries were active in steroidogenesis. Final confirmation of the role of the ovary in steroid production in DHA treated rats was obtained by measurement of serum progesterone, 17α-hydroxyprogesterone, DHA, androstenedione, testosterone, dihydrotestosterone (DHT) and estradiol levels in rats treated with DHA for 15 days using doses of 30 mg and 15 mg/kg body wt. and either sham operated or ovariectomized on day 15 followed by 5 days of vehicle treatment (Tables 1 and 2). Discontinuation of DHA for 5 days brought about a dramatic reduction in circulating levels of DHA, androstenedione, testosterone, DHT and estradiol. Furthermore, ovariectomy resulted in a significant reduction in all the circulating steroids. These observations indicate that the ovary is active in steroidogenesis in animals rendered polycystic by androgen administration.

In order to study the direct effect of androgens on the ovary leading to follicular atresia, immature rats were hypophysectomized on day 24 of life and given 32 IU pregnant mare serum gonadotrophin (PMSG) on day 30 (148). Injection of hCG 54 hours after PMSG resulted in an average of 14.9 ± 1.2 ova

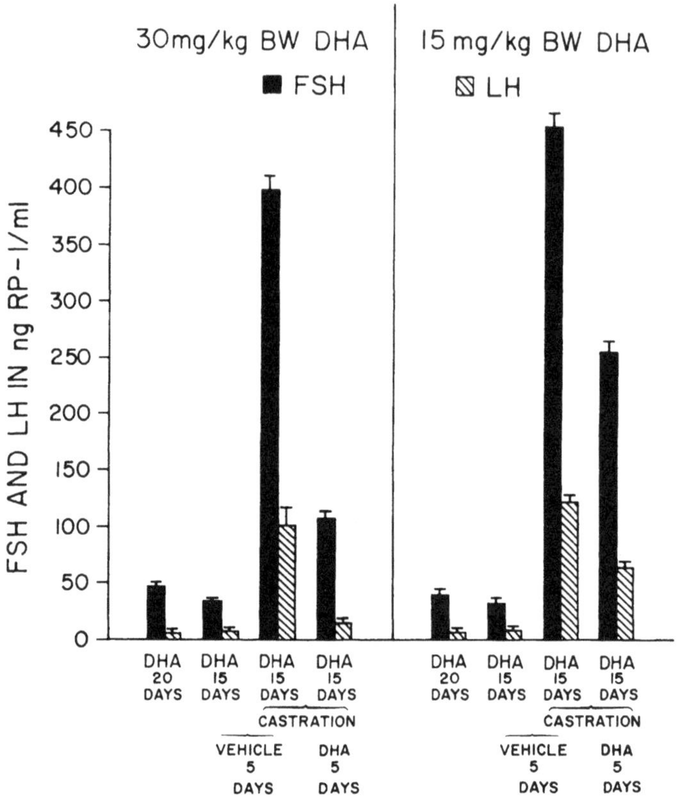

Fig. 13 Serum FSH and LH in adult female rats treated with either 3 mg/100 g b.w. or 1.5 mg/100 g b.w. DHA for 20 days or 15 days. No difference was observed in serum FSH and LH levels if DHA was administered for only 15 days followed by vehicle for 5 days. Groups of rats were castrated after 15 days of DHA administration and received either vehicle or DHA for 5 days prior to sacrifice. Marked increase in FSH and LH levels occurred after castration. Gonadotrophins remained significantly elevated [$P<0.01$] even when DHA was continued after castration compared to DHA treated intact animals. (From 14).

Table 1. Ovarian contribution to blood steroids 5 days after withdrawal of 30 mg DHA/kg body wt. given for 15 days

	Steroids in ng/ml serum		
	DHA - 15 days Vehicle 5 days	DHA - 15 days Castration Vehicle 5 days	Difference (Ovarian influence)
Progesterone	1.50 ± 0.24	0.46 ± 0.06	-1.04 ng/ml
17α-Hydroxyprogesterone	1.10 ± 0.26	0.13 ± 0.02	-0.97 ng/ml
DHA	1.54 ± 0.21	0.68 ± 0.11	-0.86 ng/ml
Androstenedione	0.16 ± 0.04	0.05 ± 0.005	-0.11 ng/ml
Testosterone	0.32 ± 0.05	0.18 ± 0.03	-0.14 ng/ml
DHT	0.17 ± 0.02	0.03 ± 0.002	-0.14 ng/ml
Estradiol	0.030 ± 0.002	0.013 ± 0.002	-0.017 ng/ml

Table 2. Ovarian contribution to blood steroids 5 days after withdrawal of 15 mg DHA/kg body wt. given for 15 days

	Steroids in ng/ml serum		
	DHA - 15 days Vehicle 5 days	DHA - 15 days Castration Vehicle 5 days	Difference (Ovarian influence)
Progesterone	1.04 ± 0.18	0.33 ± 0.08	-0.71 ng/ml
17α-Hydroxyprogesterone	0.55 ± 0.02	0.11 ± 0.01	-0.44 ng/ml
DHA	1.30 ± 0.32	0.77 ± 0.09	-0.53 ng/ml
Androstenedione	0.14 ± 0.02	0.03 ± 0.01	-0.11 ng/ml
Testosterone	0.24 ± 0.02	0.11 ± 0.04	-0.13 ng/ml
DHT	0.23 ± 0.01	0.02 ± 0.002	-0.21 ng/ml
Estradiol	0.044 ± 0.009	0.011 ± 0.001	-0.033 ng/ml

per ovulating rat with 89% of the animals ovulating. This compared with 80% ovulation rate in intact animals with 9.6 ± 1.4 ova per ovulating rat. Thus this treatment provided normal follicular development during the 54 hour period after PMSG. When various doses of the non-aromatizable androgen 5α-dihydrotestosterone (DHT) ranging from 25 μg to 400 μg/100 g body wt. were administered 30 hours after PMSG, there was a significant reduction in the percentage of healthy primary follicles (Fig. 14).

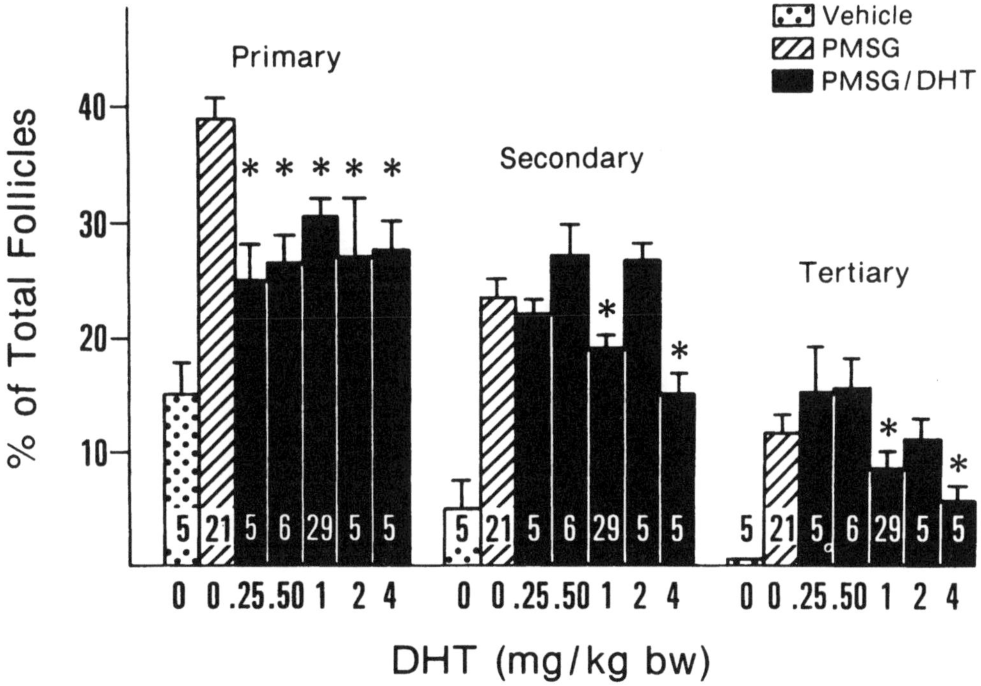

Fig. 14 Percentage of primary, secondary and tertiary ovarian follicles in rats hypophysectomized on day 24 of life and given either no treatment or 32 IU PMSG on day 30 of life. PMSG treated rats received either vehicle or various doses of DHT as a single injection, 30 h after PMSG injection and were killed 24 h later [54 h after PMSG] for the study of ovarian histology. * Indicates significant differences from the PMSG controls and the numbers at the base of the bar indicate the number of animals used. (From 148).

The percentage of atretic follicles also increased significantly (Fig. 15). These effects occurred with an exposure to DHT over a short period of time, a situation created by the single injection of a steroid that has a rapid clearance rate. To further confirm that the androgen induced follicular atresia resulted in interference with ovulation, hypophysectomized rats

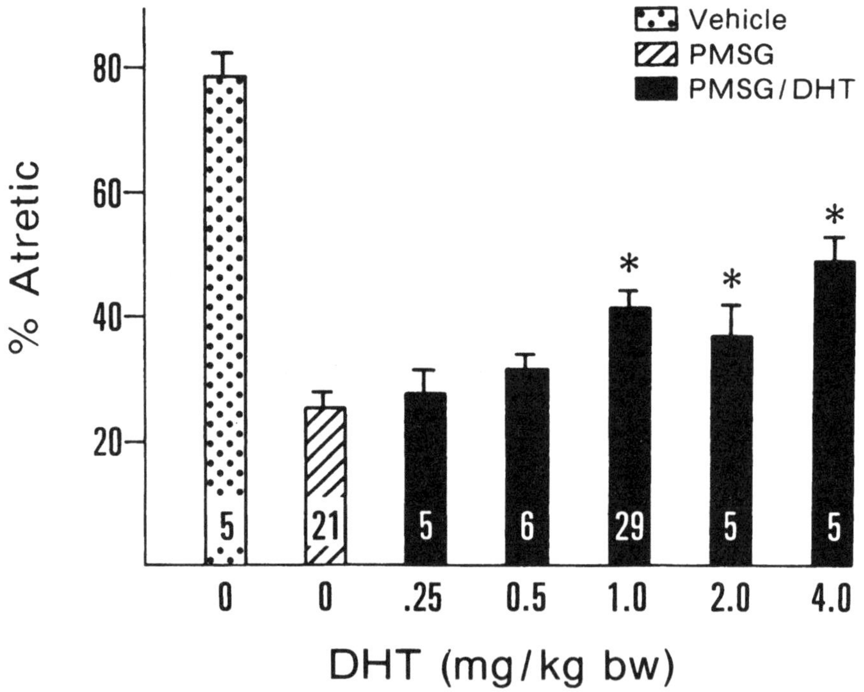

Fig. 15 Percentage of atretic follicles after PMSG or PMSG and DHT treatment in hypophysectomized female rats. Note the remarkable decrease in atresia induced by PMSG treatment and increase in atresia caused by the 1, 2 and 4 mg DHT/kg body wt. (Refer to Fig. 14 for other experimental details - From 148).

given PMSG and 1 mg DHT/kg body wt. were given 10 IU hCG, 54 hours after the initial PMSG injection. A significant reduction in the number of ova per ovulating rat occurred (Table 3). When two injections of DHT were given, one at the time of the PMSG injection and the other 30 hours later, the effect of the androgen on the reduction of the number of ova was even more dramatic.

One of the major changes brought about by DHT in the ovarian follicle should be a change in the androgen to estrogen ratio. A low androgen to estrogen ratio is considered an important criteria for healthy follicular growth (149). Primary follicles that contain only small quantities of estrogens appeared to be the most sensitive to DHT. A decrease in the number and percentage of healthy primary follicles occurred even at doses of 0.25 mg DHT/kg body wt. It was therefore of interest to examine whether administration of estradiol would protect against the atretogenic effect of DHT. Estradiol (0.25, 1 and 2 mg/kg body wt) was administered to rats at the times of PMSG and DHT injections. All doses of estradiol restored the percentage of healthy primary, secondary and tertiary follicles to those found in the PMSG controls (Fig. 16). Furthermore, the percentage of atretic follicles was significantly reduced in the DHT treated animals by all doses of estradiol (Fig. 17). When hCG was administered to these animals the DHT-induced reduction in the number of ova per ovulating rat was also reversed (Table 3) thus confirming the protective effect of estradiol on DHT induced follicular atresia.

The design of the above experiments exposed the developing ovarian follicle to the non-aromatizable androgen DHT for only short periods of time. In spite of this a considerable amount of follicular atresia took place. McNatty and coworkers (149) have described the microenvironment of antral follicles required for their progression to dominant follicles capable of ovulation in the human. Two of the critical requirements are a large number of granulosa cells that exceed 50% of those that can be present for a particular follicular diameter and a high

Table 3. Effect of steroids on ovulation induced by hCG in PMSG-primed hypophysectomized, immature rats [a]

Treatment	Rats ovulating/ injected(% rats ovulating)	Number of ova per ovulating rat (mean ± SEM)
PMSG (32 IU)	41/46 (89.1)	14.9 ± 1.2
PMSG/DHT (1 mg/kg b.w.)	25/31 (80.6)	7.8 ± 1.2[b]
PMSG/DHT (1 mg/kg b.w.; 2 injections)	10/14 (71.4)	3.2 ± 1.4[b]
PMSG/E_2 (1 mg/kg b.w.)	11/13 (84.5)	13.4 ± 2.9
PMSG/DHT/E_2 (0.25 mg/kg b.w.)	18/24 (72)	11.5 ± 2.4[c]
PMSG/DHT/E_2 (1 mg/kg b.w.)	19/23 (83)	20.5 ± 3.3[d]

[a] Rats were treated with 32 IU PMSG. DHT was given 30 h after PMSG except when two injections were given. These two injections as well as estradiol injections were given at the time of the PMSG injection and another 30 h later. hCG (10 IU, i.p.) was administered 24 h after steroid or vehicle injection. Oviducts were examined 24 h following hCG and numbers of ova counted.

[b] Significantly differs from PMSG control $P<.01$

[c] Significantly differs from PMSG control $P<.05$

[d] Significantly differs from PMSG/DHT treated $P<.01$

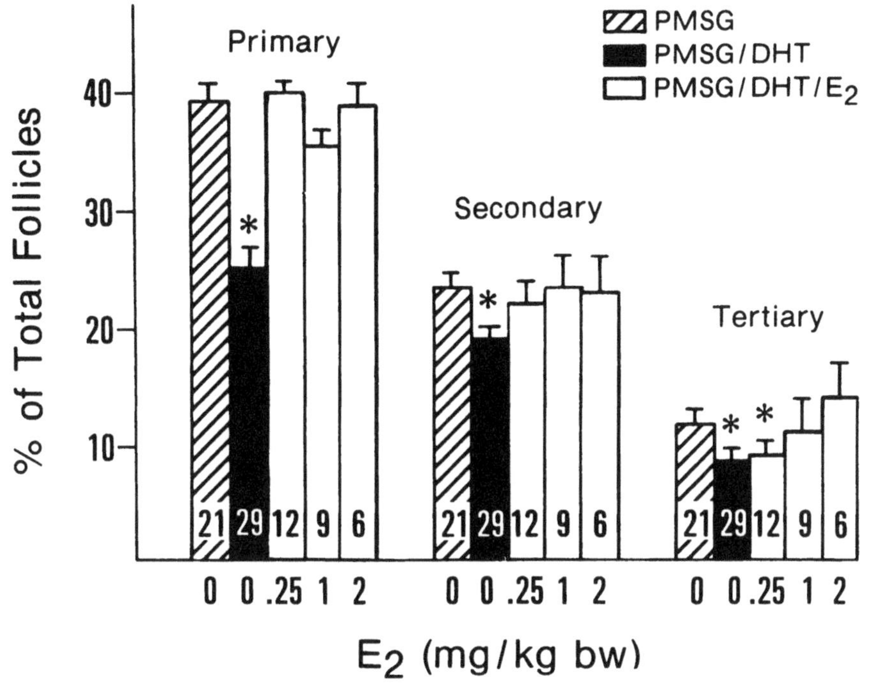

Fig. 16 Reduction in the percentage of healthy follicles by DHT prevented by the injection of various doses of estradiol at the time of PMSG injection and another injection 30 h later. (Refer to Fig. 14 for other experimental details - From 148).

estrogen to androgen ratio. The role of estradiol and FSH in the growth and development of the follicle and the generation of LH receptors has also been clearly outlined by Richards et al. (150). Administration of DHT may not only decrease the estrogen to androgen ratio but also inhibit ovarian aromatase to lower estrogen production (151). The result is an arrest of the normal growth pattern of granulosa cells leading to follicular atresia and perhaps abnormal ovarian steroidogenesis.

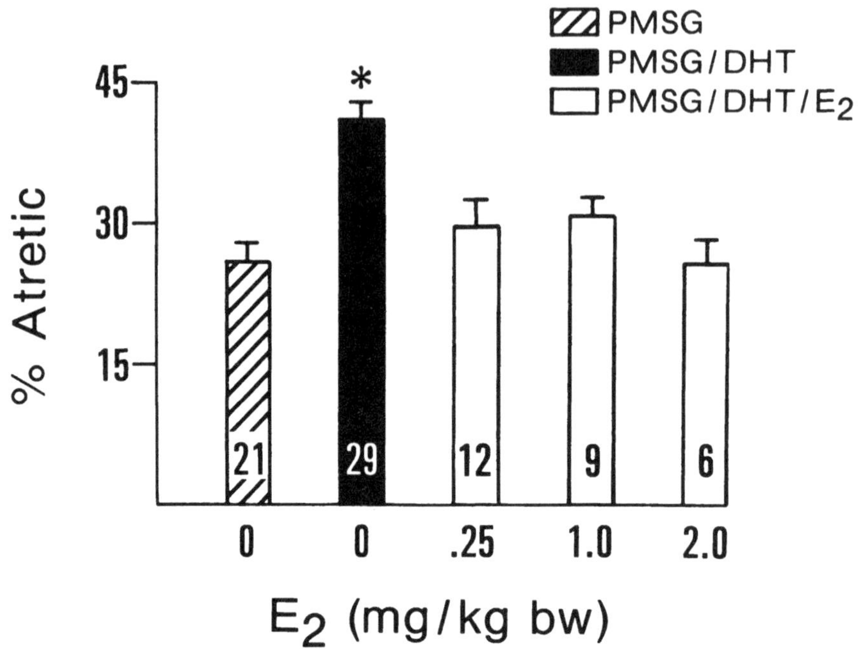

Fig. 17 Administration of two injections of estradiol as specified in Fig. 16 prevented the DHT-induced increase in the number of atretic follicles. (From 148).

Fig. 18 shows the serum steroids in hypophysectomized female rats injected with 32 IU PMSG followed by 10 IU hCG, 54 hours later. Twenty-two hours after the hCG injection the rats that failed to ovulate (1 ova or no ova in the oviducts) with treatment showed small elevations in serum androstenedione and testosterone and depressed levels of estradiol compared to rats that ovulated (Mills and Mahesh, unpublished results). The injection of DHT, 30 hours after PMSG, induced no difference in the blood steroids of rats that ovulated even though the number of ovulations were reduced (Table 3). However, the anovulatory rats showed marked increases in circulating andro-

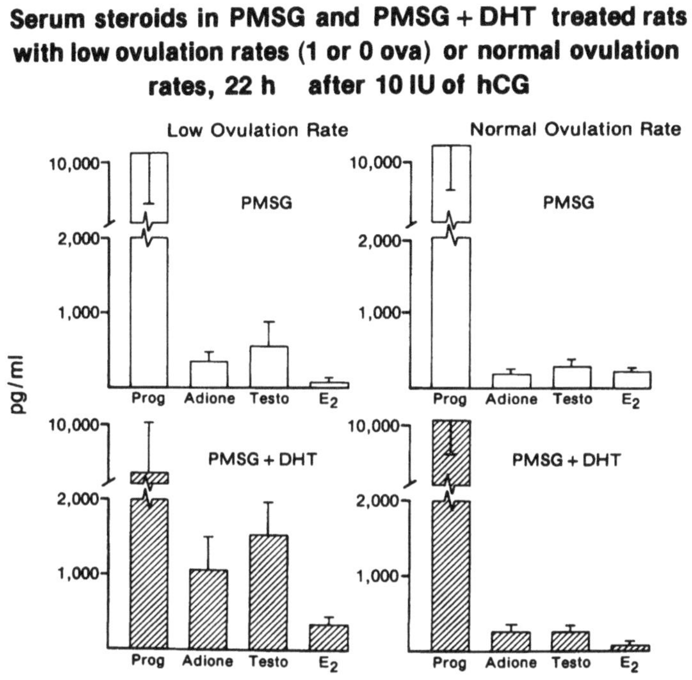

Fig. 18 Hypophysectomized rats treated with PMSG or PMSG + DHT (see Fig. 14) were administered 10 IU hCG, 54 h after PMSG. Steroids were measured in serum after 22 h. Rats were classified as low or no ovulators if ova in oviduct were absent or one ova was present. Normal ovulation rate was 6-15 ova per ovulating rat. (From Mills and Mahesh, unpublished data).

stenedione, testosterone and estradiol compared to anovulatory rats not treated with DHT. Thus androgen-induced alterations in ovarian steroidogenesis in anovulating rats is strongly indicated by these results. The granulosa cells that survive the androgen effect appear to have normal steroidogenic potential (Fig. 19) a finding that has been observed repeatedly in

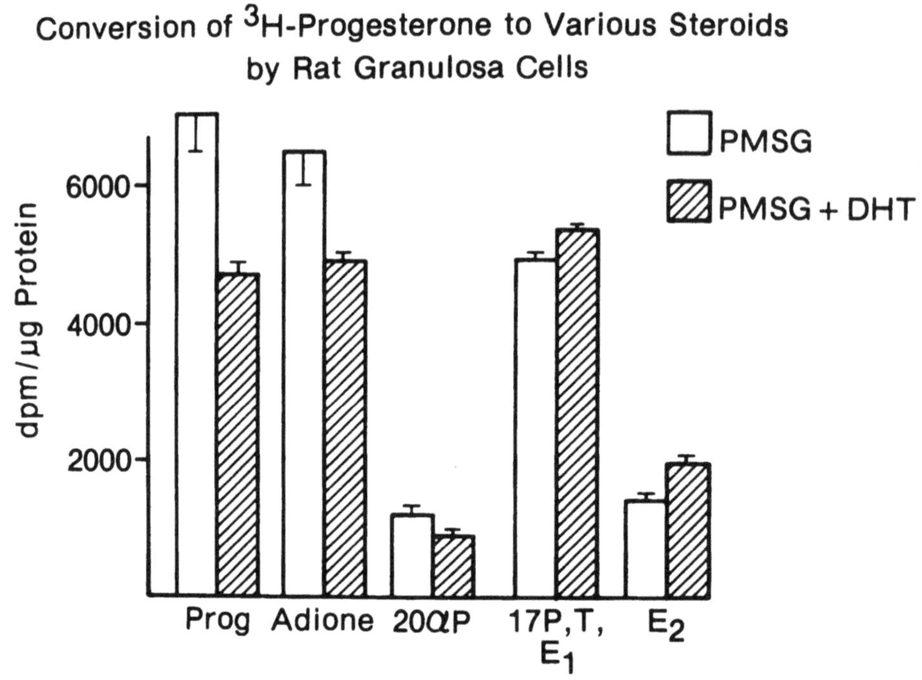

Fig. 19 No differences in the conversion of ^{3}H-progesterone to various steroids in granulosa cells from PMSG treated or PMSG + DHT treated hypophysectomized rats, 54 h after PMSG injection. (Refer to Fig. 14 for details of treatment - From Mills and Mahesh, unpublished data).

the human (138, 139). These experiments suggest the absence of inherent enzymatic abnormalities in anovulatory polycystic ovaries. The observed abnormalities, therefore, appear to be circumstantial brought about by the steroid-induced inappropriate gonadotrophin secretion and direct androgen effects on the ovary.

Summary and Conclusions

The polycystic ovary syndrome is a vaguely defined term with considerable differences in clinical and anatomical findings. Therefore, it is improbable that it can be explained on the basis of a single disorder. In spite of the fact that there is considerable evidence of the genetic transmission of the disorder, it appears that only a small percentage of patients fall into this category. Excessive androgens can arise from the adrenal and/or the ovary in this syndrome. In a number of cases this has been explained on the basis of a mild degree of failure of the adrenal 21-hydroxylase, 11β-hydroxylase and 3β-hydroxysteroid dehydrogenase. Inappropriate gonadotrophin secretion in the polycystic ovary syndrome appears to be a consequence of abnormal androgen secretion and peripheral interconversion to estrogens. Enzymatic abnormalities in steroidogenesis in the polycystic ovary have been demonstrated by several investigators on the basis of steroid content of ovarian tissue, ovarian venous blood, and cystic fluid and abnormal *in vitro* steroid biosynthesis. These observations have been confirmed by histochemical studies. In the majority of cases, however, these abnormalities are circumstantial to steroid-induced inappropriate gonadotrophin secretion and intra-ovarian androgen effects rather than being an inherent enzymatic failure within the ovary

Acknowledgement

This investigation was supported by Research Grant HD 16396 from NICHHD, NIH, U.S. Public Health Service, Bethesda, Maryland, U.S.A.

References

1. Stein, I.F., Leventhal, M.L.: Amenorrhea associated with bilateral polycystic ovaries. Am. J. Obstet. Gynecol. 29, 181-186 (1935).
2. Mahesh, V.B., Greenblatt, R.B., Aydar, C.K., Roy, S.: Secretion of androgens by the polycystic ovary and its significance. Fertil. Steril. 13, 513-530 (1962).
3. Goldzieher, J.W., Axelrod, L.R.: Clinical and biochemical features of polycystic ovarian disease. Fertil. Steril. 14, 631-641 (1963).
4. Mahesh, V.B., Greenblatt, R.B.: Steroid secretions in the normal and polycystic ovary. Recent Prog. Horm. Res. 20, 341-394 (1964).
5. Mahesh, V.B.: Steroid secretions of polycystic ovaries. Excerpta Med. Int. Congr. Ser. 83, 944-949 (1964).
6. Mahesh, V.B.: Androgen secretion and virilization. Excerpta Med. Int. Congr. Ser. 112, 213-217 (1965).
7. Shearman, R.P., Cox, R.I.: The enigmatic polycystic ovary. Obstet. Gynecol. Surv. 24, 1-33 (1966).
8. Mahesh, V.B., Greenblatt, R.B.: Adrenal and ovarian variants of the Stein-Leventhal Syndrome. Res. Steroids 3, 149-257 (1967).
9. Mahesh, V.B., Greenblatt, R.B., Scholer, H.F.L., Ellegood, J.O.: Steroid and gonadotropin secretion in the polycystic ovary syndrome. Excerpta Med. Int. Congr. Ser. 238, 160-167 (1970).
10. Kirschner, M.A., Bardin, C.W.: Androgen production and metabolism in normal virilized women. Metabolism 21, 667-688 (1972).
11. Greenblatt, R.B., Mahesh, V.B.: The androgenic polycystic ovary. Am. J. Obstet. Gynecol. 125, 712-726 (1976).
12. Parker, C.R., Mahesh, V.B.: Interrelationship between excessive levels of circulating androgens in blood and ovulatory failure. J. Reprod. Med. 17, 75-90 (1976).
13. Rebar, R., Judd, H.L., Yen, S.S.S., Rakoff, J., Vanderberg, G., Naftolin, F.: Characterization of the inappropriate gonadotropin secretion in polycystic ovary syndrome. J. clin. Invest. 57, 1320-1329 (1976).
14. Mahesh, V.B.: Current concepts of the pathophysiology of the polycystic ovary syndrome. In: "Endocrine Physiopathology of the Ovary", Eds. Tozzine, R.I., Reeves, G., Pineda, R.L., Elsevier/North-Holland Biomedical Press, Amsterdam, pp. 275-295 (1980).

15. Yen, S.S.C.: The polycystic ovary syndrome. Clin. Endocr. 12, 177-208 (1980).

16. Goldzieher, J.W.: Polycystic ovarian disease. Fertil. Steril. 25, 371-394 (1981).

17. Goldzieher, J.W., Green, J.A.: The polycystic ovary. I. Clinical and histological features. J. clin. Endocrinol. Metab. 22, 325-338 (1962).

18. Smith, K.D., Steinberger, E., Perloff, W.H.: Polycystic ovarian disease. Am. J. Obstet. Gynecol. 93, 994-1001 (1965).

19. Cervino, J.M., Ravera, J.J., Grosso, O.F., Pou de Santigo, A., Morato Manaro, J., Mussio Fournier, J.C.: Stein-Leventhal Syndrome: two sisters attacked by this syndrome. Ann. Endocrinol. 17, 355-368 (1956).

20. Evans, T.N., Riley, G.M.: Polycystic ovarian disease: etiology and rationale for surgical treatment. Obstet. Gynecol. 12, 168-178 (1958).

21. Trace, R.J., Keaty, E.C., McCall, M.L.: An investigation of ovarian tissue and urinary 17-ketosteroids in patients with bilateral polycystic ovaries. Am. J. Obstet. Gynecol. 79, 310-315 (1960).

22. Mahesh, V.B., Toledo, S.P.A., Mattar, E.: Hormone levels following wedge-resection in polycystic ovary syndrome. Obstet. Gynecol. 51, 64S-69S (1978).

23. Jeffcoate, T.N.: The androgenic ovary with special reference to the Stein-Leventhal Syndrome. Am. J. Obstet. Gynecol. 88, 143-156 (1964).

24. McDonough, P.G., Mahesh, V.B., Ellegood, J.O.: Steroids, FSH and LH profiles in identical twins with polycystic ovaries. Am. J. Obstet. Gynecol. 113, 1072-1078 (1972).

25. Suzor, R., Falloux, P.: Is the Stein-Leventhal Syndrome hereditary. Bull. Fed. Gynecol. Obstet. (France) 13, 12-13 (1961).

26. Castlemen, B., Kibbee.: Case records of the Massachusetts General Hospital. New Engl. J. Med. 257, 1086-1093 (1957).

27. Vague, J., Picard, D., Vitry, G., Galinier, L., Favier, G., Miller, G.: Fréquence des intersexualités somatiques aux voies génitales normales. Ann. Endocrinol. (Paris) 17, 485-489 (1956).

28. Wais, S.: Bilateral polycystic ovaries with negative chromatin pattern. Acta Endocrinol. 51 (suppl.) abstract 615 (1960).

29. Netter, A., Bloch-Michel, H., Salomon, Y., Thervet, F., de Grouchy, J., Lamy, M.: Study of karyotypes in Stein-Leventhal Syndrome. Ann. Endocrinol. (Paris) 22, 841-849 (1961).

30. Short, R.B.: Further observations on the defective synthesis of ovarian steroids in the Stein-Leventhal Syndrome. J. Endocrinol. 24, 359-365 (1962).

31. Byrd, J.R., Mahesh, V.B., Greenblatt, R.B.: Chromosomal studies in the Stein-Leventhal Syndrome. J. clin. Endocrinol. Metab. 24, 939-940 (1964).

32. Cooper, H.E., Spellacy, W.N., Prem, K.A., Cohen, W.D.: Hereditary factors in the Stein-Leventhal Syndrome. Am. J. Obstet. Gynecol. 100, 371-387 (1968).

33. Givens, J.R., Wiser, W.L., Coleman, S.A., Wilroy, R.S., Anderson, R.N., Fish, S.A.: Familial ovarian hyperthecosis. Am. J. Obstet. Gynecol. 110, 959-972 (1971).

34. Cohen, P.N., Givens, J.R., Wiser, W.L., Wilroy, R.S., Summitt, R.L., Coleman, S.A., Anderson, R.N.: Polycystic ovarian disease, maturation arrest of spermiogenesis and Klinefelter's Syndrome in siblings of a family with familial hirsutism. Fertil. Steril. 26, 1228-1238 (1975).

35. Wilroy, R.S., Givens, J.R., Wiser, W.L., Coleman, S.A., Anderson, R.A., Fish, S.A.: Hyperthecosis - an inheritable form of polycystic ovarian disease. In: "Genetic Forms of Hypogonadism", Ed. Bergsma, D., Birth Defects 11, 81-85 (1975).

36. Parker, R., Ming, P.L., Rajan, R., Goodner, D.M., Reme, G.: Clinical and cytogenetic studies of patients with polycystic ovarian disease. Am. J. Obstet. Gynecol. 137, 656-660 (1980).

37. Buxton, C.L., Vande Wiele, R.L.: Wedge-resection for polycystic ovaries. New Engl. J. Med. 251, 293-297 (1954).

38. Perloff, W.H., Hadd, H.E., Channick, B.J., Nodine, J.H.: Hirsutism. Arch. Intern. Med. 100, 981-985 (1957).

39. Gallagher, T.F., Kappas, A., Hellman, L., Lipsett, M.B., Pearson, O.H., West, C.D.: Adrenal hyperfunction of idiopathic hirsutism and Stein-Leventhal Syndrome. J. clin. Invest. 37, 794-799 (1958).

40. Brooks, R.V., Prunty, F.T.G.: Patterns of steroid excretion in three types of post-pubertal hirsutism. J. Endocrinol. 21, 263-276 (1960).

41. Lipsett, M.B., Riter, B.: Urinary ketosteroids and pregnanetriol in hirsutism. J. clin. Endocrinol. Metab. 20, 180-186 (1960).

42. Kase, N., Kowal, J., Perloff, W., Soffer, L.J.: *In vitro* production of androgens by a virilizing adrenal adenoma and associated with polycystic ovaries. Acta Endocrinol. 44, 15-19 (1963).

43. Giest, S.H., Gaines, J.A.: Diffuse luteinization of the ovaries associated with the masculinization syndrome. Am. J. Obstet. Gynecol. 43, 975-983 (1942).

44. Feher, L., Gijory, G., Less, E., Laszlo, G.: Gleichzeitiges auftreten von pseudohermaphroditismus femininus und hyperthecosis ovarii bei geschwistern. Acta Endocrinol. 28, 219-226 (1958).

45. Perloff, W.H., Channick, B.J., Hadd, H.E., Nodine, J.H.: The Stein-Leventhal Ovary. Fertil. Steril. 9, 247-255 (1958).

46. Greenblatt, R.B., Martinez-Manautou, J., Clark, S.L., Rosenberg, A.P.: Suppression of adrenal cortical activity in treatment of menstrual disorders. Metabolism 7, 25-39 (1958).

47. Kovacic, N.: Congenital adrenal hyperplasia and precocious gonadotropin resection in a 6-year-old girl. J. clin. Endocrinol. Metab. 19, 844-847 (1959).

48. Perloff, W.H., Channick, B.J., Suplick, B., Carrington, E.R.: Clinical management of idiopathic hirsutism. J. Am. Med. Assoc. 167, 2041-2047 (1958).

49. Bergman, P., Sjogren, B., Hakansson, B.: Hypertensive form of congenital adrenal hyperplasia. Acta Endocrinol. 40, 555-564 (1962).

50. Gabrilove, J.L., Sharma, D.C., Dorfman, R.I.: Adrenocortical 11β-hydroxylase deficiency and virilism first manifested in an adult woman. New Engl. J. Med. 272, 1189-1194 (1965).

51. Greenblatt, R.B.: Cortisone in the treatment of the hirsute woman. Am. J. Obstet. Gynecol. 66, 700-710 (1953).

52. Gemzell, C.A., Tillinger, K.G., Westman, A.: A clinical study of ovarian pathology in the urinary excretion of 17-ketosteroids. Acta Endocrinol. 30, 387-404 (1959).

53. Pesonen, S., Timonen, S., Mikkonen, R.: Symptoms and etiology of the Stein-Leventhal Syndrome. Acta Endocrinol. 30, 405-423 (1959).

54. Mahesh, V.B., Greenblatt, R.B.: Physiology and pathogenesis of the Stein-Leventhal Syndrome. Nature 191, 888-890 (1961).

55. Mahesh, V.B., Greenblatt, R.B.: Isolation of dehydroepiandrosterone and 17α-hydroxy-Δ^5-pregnenolone from the polycystic ovaries of Stein-Leventhal Syndrome. J. clin. Endocrinol. Metab. 22, 441-448 (1962).

56. Mahesh, V.B., Greenblatt, R.B., Aydar, C.K., Roy, S., Puebla, R.A., Ellegood, J.O.: Urinary steroid patterns in hirsutism. I. J. clin. Endocrinol. Metab. 24, 1283-1292 (1964).

57. Mahesh, V.B., Greenblatt, R.B.: Urinary steroid excretion patterns in hirsutism. II. J. clin. Endocrinol. Metab. 24, 1293-1302 (1964).

58. Starka, L., Matys, Z., Janata, J.: Der nachweis von dehydroepiandrosterone in menschlichen sklerocystischen ovarien. Clin. chim. Acta 7, 776-779 (1962).

59. Baulieu, E.E., Mauvais-Jarvis, P., Corpechot, C.: Steroid studies in a case of Stein-Leventhal Syndrome with hirsutism. J. clin. Endocrinol. Metab. 23, 374-382 (1963).

60. Short, R.V., London, D.R.: Defective biosynthesis of ovarian steroids in the Stein-Leventhal Syndrome. Br. med. J. 1, 1724-1727 (1961).

61. Mahajan, D.K., Shah, P.N., Eik-Nes, K.B.: Steroids in cyst fluid obtained from polycystic ovaries in hirsute women. J. Obstet. Gynaec. Br. Commonw. 70, 8-12 (1963).

62. Parker, C.R., Bruneteau, D.W., Greenblatt, R.B., Mahesh, V.B.: Peripheral ovarian and adrenal vein steroids in hirsute women: Acute effects of hCG and ACTH. Fertil. Steril. 26, 877-888 (1975).

63. Gold, J.G., Scommegna, A.: Use of corticoids in obstetrics and gynecology. Clin. Obstet. Gynecol. 3, 1068-1082 (1960).

64. Nichols, T., Nugent, C.A., Tyler, F.H.: Glucocorticoid suppression of urinary testosterone excretion in patients with idiopathic hirsutism. J. clin. Endocrinol. Metab. 26, 79-86 (1966).

65. Abraham, G.E., Marovlis, G.B., Buster, J.E., Chang, R.J., Marshall, J.R.: Effects of dexamethasone on serum cortisol and androgen levels in hirsute patients. Obstet. Gynecol. 47, 395-402 (1976).

66. Nichols, T., Nugent, C.A., Tyler, F.H.: Diurnal variation in suppression of adrenal function by glucocorticoids. J. clin. Endocrinol. Metab. 25, 343-349 (1965).

67. Lopez, J.M., Migeon, C.J., Jones, G.E.S.: Hirsutism and evaluation of the dexamethasone suppression and chorionic stimulation test. Am. J. Obstet. Gynecol. 98, 749-758 (1967).

68. Bardin, C.W., Hembree, W.C., Lipsett, M.B.: Suppression of testosterone and androstenedione production rates with dexamethasone in women with idiopathic hirsutism and polycystic ovaries. J. clin. Endocrinol. Metab. 28, 1300-1306 (1968).

69. Cruikshank, D.P., Chapler, F.K., Yannone, M.E.: Differential adrenal and ovarian suppression. Obstet. Gynecol. 38, 724-730 (1971).

70. Cooke, C.W., McEvoy, D., Bulaschenko, H., Wallach, E.E.: Adrenocortical and ovarian function in the hirsute woman. Am. J. Obstet. Gynecol. 114, 65-77 (1972).

71. Ettinger, B., Goldfield, E.B., Burrill, R.C., Von Werder, K., Forsham, P.H.: Plasma testosterone stimulation-suppression dynamics in hirsute women. Am. J. Med. 54, 195-200 (1973).

72. Polansky, S.: Luteinizing hormone, adrenal androgenesis and polycystic ovarian disease. Obstet. Gynecol. 45, 451-454 (1974).

73. Givens, J.R., Andersen, R.N., Ragland, J.B., Wisert, W.L., Umstot, E.S.: Adrenal function in hirsutism. J. clin. Endocrinol. Metab. 40, 988-1000 (1975).

74. Cunningham, G.R., Goldzieher, J.W., de la Pena, A., Oliver, M.: The mechanism of ovulation induction by triamcinolone acetonide. J. clin. Endocrinol. Metab. 46, 8-14 (1978).

75. Kirschner, A., Jacobs, J.: Combined ovarian and adrenal vein catheterization to determine the sites of androgen overproduction in hirsute women. J. clin. Endocrinol. Metab. 33, 199-209 (1971).

76. Rosenfield, R.L., Ehrlich, E.N., Cleary, R.E.: Adrenal and ovarian contributions to the elevated free plasma androgen levels in hirsute women. J. clin. Endocrinol. Metab. 34, 92-98 (1972).

77. Benjamin, F., Cohen, M., Romney, S.L.: Sequential adrenal and ovarian suppression tests in the differential diagnosis of the polycystic ovary syndrome. Fertil. Steril. 21, 854-859 (1970).

78. Beck, W.W., McEvoy, D., Wallach, E.E.: Combined dexamethasone and human chorionic gonadotropin in diagnosis of polycystic ovarian disease. Obstet. Gynecol. 37, 211-217 (1971).

79. Aakvaag, A., Vogt, J.H., Fylling, P.: Plasma and urinary androgens in hirsute women during adrenal and ovarian suppression. Acta Endocrinol. 64, 103-110 (1970).

80. Abraham, G.E.: Ovarian and adrenal contributions to peripheral androgens during the menstrual cycle. J. clin. Endocrinol. Metab. 39, 340-346 (1974).

81. Kim, M.H., Hosseinian, A.H., Dupont, C.: Plasma levels of estrogens, androgens and progesterone during normal and dexamethasone treated cycles. J. clin. Endocrinol. Metab. 39, 706-712 (1974).

82. Stahh, N.L., Teeslink, C.R., Greenblatt, R.B.: Ovarian adrenal and peripheral testosterone levels in the polycystic ovary syndrome. Am. J. Obstet. Gynecol. 117, 194-200 (1973).

83. Kirschner, M.A.: Adrenal and gonadal venous catheterization studies in hirsute women. In: "Hirsutism and Virilism", Eds. Mahesh, V.B., Greenblatt, R.B., Wright-PSG Press, pp. 309-332 (1983).

84. Ingersoll, R.M., McDermott, W.B.: Bilateral polycystic ovaries. Stein-Leventhal Syndrome. Am. J. Obstet. Gynecol. 60, 117-125 (1950).

85. McArthur, J.W., Ingersoll, R.M., Worcester, J.: The urinary excretion of ICSH and FSH by women with diseases of the reproductive system. J. clin. Endocrinol. Metab. 18, 1202-1215 (1958).

86. Taymor, M.L., Barnard, R.: Luteinizing hormone excretion in the polycystic ovary syndrome. Fertil. Steril. 13, 501-512 (1962).

87. Yen, S.S.C., Vela, P., Rankin, J.: Inappropriate secretion of follicle stimulating hormone and luteinizing hormone in polycystic ovarian disease. J. clin. Endocrinol. Metab. 30, 435-442 (1970).

88. Gambrell, R.D., Greenblatt, R.B., Mahesh, V.B.: Serum gonadotropin levels and ancillary studies in Stein-Leventhal Syndrome treated with clomiphene citrate. Obstet. Gynecol. 38, 850-862 (1971).

89. Gambrell, R.D., Greenblatt, R.B., Mahesh, V.B.: Inappropriate secretion of LH in the Stein-Leventhal Syndrome. Obstet. Gynecol. 42, 429-440 (1973).

90. Berger, M.J., Taymor, M.L., Patton, W.C.: Gonadotropin levels and secretory patterns with typical and atypical polycystic ovarian disease. Fertil. Steril. 26, 619-626 (1975).

91. De Vane, G.W., Czekala, N.M., Judd, H.L., Yen, S.S.C.: Circulating gonadotropins, estrogens and androgens in polycystic ovarian disease. Am. J. Obstet. Gynecol. 121, 496-500 (1975).

92. Duignan, N.M., Shaw, R.W., Rudd, B.T., Holder, G., Williams, J.W., Butt, W.R., Logan Edwards, R., London, D.R.: Sex hormone levels and gonadotropin release in the polycystic ovary syndrome. Clin. Endocr. 4, 287-295 (1975).

93. Baird, D.T., Corker, D.T., Davidson, D.W., Hunter, H.M., Michie, E.A., Van Look, P.F.A.: Pituitary ovarian relationships in polycystic ovary syndrome. J. clin. Endocrinol. Metab. 45, 798-809 (1977).

94. Patton, W.C., Berger, M.J., Thompson, I.E., Cheng, A.P., Grimes, E.M., Taymor, M.L.: Pituitary gonadotropin responses to synthetic LHRH in patients with typical and atypical polycystic ovarian disease. Am. J. Obstet. Gynecol. 121, 382-386 (1975).

95. Taymor, M.L., Thompson, I.E., Berger, M.J., Patton, W.C.: LHRH as a diagnostic and research tool in gynecological endocrinology. Am. J. Obstet. Gynecol. 120, 721-732 (1974).

96. Oettinger, V., Paroudakis, A., Mahesh, V.B., Greenblatt, R.B.: Effects of LHRH stimulation in anovulatory women. Obstet. Gynecol. 45, 614-618 (1975).

97. Moltz, L., Rommler, A., Schwartz, V., Bidlingmaier, F., Hammerstein, J.: Peripheral steroid-gonadotropin interactions and diagnostic significance of double stimulation tests with LHRH in polycystic ovarian disease. Am. J. Obstet. Gynecol. 134, 813-818 (1979).

98. Raj, S.G., Thompson, I.E., Berger, M.J., Taymor, M.L.: Clinical aspects of polycystic ovary syndrome. Obstet. Gynecol. 49, 552-556 (1977).

99. Givens, J.R., Anderson, R.N., Umstot, E.S., Wiser, W.L.: Clinical findings and hormonal responses in patients with polycystic ovarian disease with normal versus elevated LH levels. Obstet. Gynecol. 47, 388-394 (1976).

100. Rebar, R.W., Harman, S.M., Vaitukaitis, J.L.: Differential responsiveness to LRF after estrogen therapy in women with hypothalamic amenorrhea. J. clin. Endocrinol. Metab. 46, 48-54 (1978).

101. Yen, S.S.C., Vela, P., Ryan, K.J.: Effect of clomiphene citrate in the polycystic ovary syndrome: Relationship between serum gonadotropin and corpus luteum function. J. clin. Endocrinol. Metab. 31, 7-13 (1970).

102. Yen, S.S.C., Lasley, B.L., Wang, C.F., Leblanc, H., Siler, T.M.: The operating characteristics of the hypothalamic-pituitary system during the menstrual cycle and observations of biological actions of somatostatin. Recent Prog. Horm. Res. 31, 321-363 (1975).

103. Nillus, S.J., Wide, L.: Induction of a midcycle-like peak of LH in young women by exogenous estradiol. J. Obstet. Gynaec. Br. Commonw. 78, 822-827 (1972).

104. Thomas, K., Cardon, M., Donnez, J., Ferrin, J.: Changes in hypophyseal responsiveness to synthetic LHRH during the normal menstrual cycle in women. Contraception 7, 289-297 (1973).

105. Yen, S.S.C., Vandenberg, G., Rebar, R., Ehara, Y.: Variation of pituitary responsiveness to synthetic LRF during different phases of the menstrual cycle. J. clin. Endocrinol. Metab. 35, 931-937 (1972).

106. Jaffe, R.B., Keye, W.R.: Estradiol augmentation of pituitary responsiveness to gonadotropin releasing hormone in women. J. clin. Endocrinol. Metab. 39, 850-855 (1974).

107. Yen, S.S.C., Vandenberg, G., Siler, T.M.: Modulation of pituitary responsiveness to LRF by estrogen. J. clin. Endocrinol. Metab. 39, 170-177 (1974).

108. Mortimer, R.H., Lev-Gur, M., Freeman, R., Fleischer, N.: Pituitary response to bolus and continuous intravenous infusion of LHRH in normal women and women with polycystic ovarian syndrome. Am. J. Obstet. Gynecol. 130, 630-634 (1978).

109. Katz, M., Carr, P.J., Cohen, B.M., Millar, R.P.: Hormonal effects of wedge-resection of polycystic ovaries. Obstet. Gynecol. 51, 437-444 (1978).

110. Kandeel, F.R., Butt, W.R., London, D.R., Lynch, S.S., Logan Edwards, R., Rudd, B.T.: Oestrogen amplification of LHRH response to clomiphene. Clin. Endocr. 9, 429-441 (1978).

111. Chang, R.J., Mander, F.P., Lu, J.K.H., Judd, H.L.: Enhanced disparity of gonadotropin secretion by estrone in women with polycystic ovarian disease. J. clin. Endocrinol. Metab. 54, 490-494 (1982).

112. Black, J.B., Mahesh, V.B.: Effect of dehydroepiandrosterone on pituitary gonadotropins and ovulation in the rat. Fed. Proc. 28, 506 (Abstract) (1969).

113. Knudsen, J.F., Mahesh, V.B.: Initiation of precocious sexual maturation in the immature rat treated with dehydroepiandrosterone. Endocrinology 97, 458-468 (1975).

114. Knudsen, J.F., Costoff, A., Mahesh, V.B.: Dehydroepiandrosterone induced polycystic ovaries and acyclicity in the rat. Fertil. Steril. 26, 807-817 (1975).

115. Parker, C.R., Mahesh, V.B.: Dehydroepiandrosterone induced precocious ovulation: Correlative changes in blood steroids, gonadotropins and cytosol estradiol receptors of the anterior pituitary gland and the hypothalamus. J. Steroid. Biochem. 8, 173-177 (1977).

116. Parker, C.R., Mahesh, V.B.: Hormonal events surrounding the natural onset of puberty in female rats. Biol. Reprod. 14, 347-353 (1976).

117. Parker, C.R., Costoff, A., Muldoon, T.G., Mahesh, V.B.: Actions of PMSG in the immature female rat: Correlative changes in blood steroids, gonadotropins and cytoplasmic estradiol receptors of the anterior pituitary and hypothalamus. Endocrinology 98, 129-138 (1975).

118. Ward, R.C., Costoff, A., Mahesh, V.B.: The induction of polycystic ovaries in mature cycling rats by the administration of dehydroepiandrosterone. Biol. Reprod. 18, 614-623 (1978).

119. Ikkos, D.G., Dellia-Sfikaki, A.: Plasma steroids in hirsutism. Obstet. Gynecol. 46, 114 (1975).

120. Lobo, R.A., Goebelsmann, U.: Adult manifestation of congenital adrenal hyperplasia due to incomplete 21-hydroxylase deficiency mimicking polycystic ovarian disease. Am. J. Obstet. Gynecol. 138, 720-726 (1980).

121. Lobo, R.A., Goebelsmann, U.: Evidence for reduced 3β-ol hydroxysteroid dehydrogenase activity in some hirsute women thought to have polycystic ovary syndrome. J. clin. Endocrinol. Metab. 53, 394-400 (1981).

122. Lobo, R.A., Goebelsmann, U.: Effect of androgen excess on inappropriate gonadotropin secretion as found in the polycystic ovary syndrome. Am. J. Obstet. Gynecol. 142, 394-401 (1982).

123. Blankstein, J., Faiman, C., Reyes, F.I., Schroeder, M.L., Winter, J.S.D.: Adult-onset familial adrenal 21-hydroxylase deficiency. Am. J. Med. 68, 441-448 (1980).

124. Chrousos, G.P., Loriaux, D.L., Mann, D.L., Cutler, G.B.: Late-onset 21-hydroxylase deficiency mimicking idiopathic hirsutism or polycystic ovarian disease. Ann. Int. Med. 96, 143-148 (1982).

125. Lachelin, G.C.L., Barnett, M., Hopper, B.R., Brink, G., Yen, S.S.C.: Adrenal function in normal women and women with polycystic ovary syndrome. J. clin. Endocrinol. Metab. 49, 882-898 (1979).

126. Sharma, D.C., Forchielli, E., Dorfman, R.I.: Inhibition of steroidal 11β-hydroxylation by androgens. J. biol. Chem. 238, 572-575 (1963).

127. Leventhal, M.L., Scommegna, A.: Multiglandular effects of the Stein-Leventhal Syndrome. Am. J. Obstet. Gynecol. 87, 445-454 (1963).

128. Axelrod, L.R., Goldzieher, J.W.: Enzymatic inadequacies in human polycystic ovaries. Arch Biochem. Biophys. 95, 457-548 (1961).

129. Ikonen, M., Neimi, M., Personen, S., Timonen, S.: Histochemical localization of four dehydrogenase systems in human ovary during the menstrual cycle. Acta Endocrinol. 38, 293-302 (1961).

130. Deane, W.H., Lober, B.L., Romney, S.L.: Enzymatic histochemistry of normal human ovaries of the menstrual cycle, pregnancy and the early puerperium. Am. J. Obstet. Gynecol. 83, 281-294 (1962).

131. Dokumou, S.I., Dasher, G.I.: A histochemical study of the ovaries of women suffering from the Stein-Leventhal Syndrome. Am. J. Obstet. Gynecol. 86, 183-187 (1963).

132. Goldberg, B., Seegar-Jones, G.E., Turner, D.A.: Steroid 3β-hydroxysteroid dehydrogenase in human endocrine tissue. Am. J. Obstet. Gynecol. 86, 349-359 (1963).

133. Axelrod, L.R., Goldzieher, J.W., Ross, S.D.: Concurrent 3β-hydroxysteroid dehydrogenase deficiency in adrenal and sclerocystic ovary. Acta Endocrinol. 48, 392-412 (1965).

134. Zander, J., Wiest, W.G., Ober, K.G.: Clinical, histologic and biochemical observations and polycystic ovaries with concomitant adenomatus atypical hyperplasia of the endometrium: transformation of progesterone in ovarian tissue. Arch. Gynäkologie 196, 481-503 (1962).

135. Ances, I.G., Haskins, A.L., Ganis, F.M.: Metabolism of ovarian steroids in a patient with chronic familial anovulation. Am. J. Obstet. Gynecol. 108, 1130-1133 (1970).

136. Maschler, I., Salzberger, M., Finkelstein, M.: 11β-Hydroxylase with affinity to C-21-deoxysteroids from ovaries of patients with polycystic ovary syndrome. J. clin. Endocrinol. Metab. 41, 999-1002 (1975).

137. Maschler, I., Salzberger, M., Finkelstein, M.: Ovarian enzymatic divergence in patients with polycystic ovary syndrome excreting urinary pregnanetriol. Acta Endocrinol. 82, 366-379 (1976).

138. Wilson, E.A., Erickson, G.F., Zartuski, P., Finn, A.E., Tulchinsky, D., Ryan, K.J.: Endocrine studies of normal and polycystic ovarian tissue *in vitro*. Am. J. Obstet. Gynecol. 134, 56-63 (1979).

139. McNatty, K.P., Smith, D.M., Makris, A., De Grazia, C., Tulchinsky, D., Osathanondh, R., Schiff, I., Ryand, K.J.: The intraovarian sites of androgen and estrogen formation in women with normal and hyperandrogenic ovaries as judged by *in vitro* experiments. J. clin. Endocrinol. Metab. 50, 755-763 (1980).

140. Payne, R.W., Hellbaum, A.A., Owens, J.N.Jr.: The effects of androgens on the ovaries and uterus of estrogen-treated hypophysectomized immature female rat. Endocrinology 59, 306-316 (1956).

141. Louvet, J.P., Harman, S.M., Ross, G.T.: Effects of hCG, hICSH and hFSH on ovarian weights in estrogen primed hypophysectomized immature female rats. Endocrinology 96, 1179-1186 (1975).

142. Louvet, J.P., Harman, S.M., Schreiber, J.R., Ross, G.T.: Evidence for a role of androgens in follicular maturation. Endocrinology 97, 366-372 (1975).

143. Hillier, S.G., Ross, G.T.: Effects of exogenous testosterone on ovarian weight, follicular morphology and intra-ovarian progesterone concentration in estrogen-primed hypophysectomized immature female rats. Biol. of Reprod. 20, 261-268 (1979).

144. Farooki, R.: Effects of androgen on induction of gonadotropin receptors and gonadotropin stimulated cAMP production in rat ovarian granulosa cells. Endocrinology 106, 1216-1233 (1980).

145. Schreiber, J.R., Reid, R., Ross, G.T.: A receptor-like testosterone binding protein in ovaries from estrogen stimulated hypophysectomized immature female rats. Endocrinology 98, 1206-1213 (1976).

146. Schreiber, J.R., Ross, G.T.: Further characterization of a rat ovarian testosterone receptor with evidence for nuclear translocation. Endocrinology 99, 590-596 (1976).

147. Judd, H.L., Rigg, L.A., Anderson, D.C., Yen, S.S.C.: The effects of ovarian wedge-resection on circulating gonadotropin and ovarian steroid levels in patients with polycystic ovary syndrome. J. clin. Endocrinol. Metab. 43, 347-355 (1976).

148. Bagnell, C.A., Mills, T.M., Costoff, A., Mahesh, V.B.: A model for the study of androgen effects on follicular atresia and ovulation. Biol. of Reprod. 27, 903-914 (1982).

149. McNatty, K.P., Smith, D.M., Makris, A., Osathanondh, R., Ryan, K.J.: The microenvironment of the human antral follicle. Interrelationships among the steroid levels in antral fluid, the population of granulosa cells and the status of the oocyte *in vivo* and *in vitro*. J. clin. Endocrinol. Metab. 49, 851-860 (1979).

150. Richards, J.S., Jonassen, J.A., Rolfes, A.I., Kersey, K., Reichart, L.E.: Adenosine-3'-5'-monophosphate, LH receptor and progesterone during granulosa cell differentiation. Endocrinology 104, 765-773 (1979).

151. Hillier, S.G., Van Den Boogaard, A.M.J., Reichert, L.E., van Hall, E.V.: Intraovarian sex steroid hormone interactions and the regulation of follicular maturation. J. clin. Endocrinol. Metab. 50, 640-647 (1980).

STEROID METABOLIZING ENZYMES IN HUMAN BREAST CANCER

Y. J. Abul-Hajj
Department of Pharmaceutical Cell Biology, College of Pharmacy, University of Minnesota, Minneapolis, Minnesota, U.S.A.

I. Introduction

The observation made by Beatson in 1896 (1) of the beneficial effects of castration in the treatment of breast cancer has stimulated much interest and research over the last 85 years. It is now well established that approximately one-third of women with metastatic breast cancer experience objective regression after endocrine ablative or hormone therapy. While the reasons for response to oophorectomy and hypophysectomy are not completely understood, one possible explanation is the removal of circulating estrogens from the system. However, the exact reasons for adrenalectomy were not completely understood. Furthermore, the rationale for using pharmacological doses of steroids has yet to be explained.

Steroid therapy is effective in about one-third of patients with breast cancer but only for a limited period of time. The efficacy of these compounds may be due to several factors including their concentration in blood, tissue distribution and metabolism, not only by liver and other normal cells, but also by the tumor cells. The hormonal environment in which mammary tumors develop and are maintained is important in understanding the disease. The study of steroid hormone levels in blood and urine may not be sufficient and efforts have recently been carried out to study steroid metabolism in mammary tumor tissues. These studies may be important not only in our understanding of the underlying mechanisms of hormone dependence

Steroid Converting Enzymes and Diseases

and independence but may also help to explain the different responses to steroid hormone therapy.

In this brief review, highlights of the data which has appeared in the literature over the past 20 years on the *in vitro* metabolism of steroids by human mammary tumors will be reviewed. These studies were initiated as the result of several earlier observations made by West et al. (2) and Dao and Morreal (3) who showed the presence of estrogens in urine of oophorectomized adrenalectomized patients with breast cancer maintained on androgen therapy. Thus, these studies suggested the presence of extra-gonadal and extra-adrenal synthesis of estrogens. These initial studies led to the proliferation of results reported in the early and mid 1970's on the peripheral aromatization of androgens. Thus, it is now well established that aromatization of androgens takes place in several tissues including skin (4), brain (5), liver (6), muscle (7) and adipose cells (7, 8, 9, 10). Furthermore, Siiteri and coworkers (11, 12) have shown that androstenedione, which is primarily of adrenal origin, is the immediate precursor for estrone formation in postmenopausal women. While these results played a significant role in our understanding of estrogen synthesis in postmenopausal women, they still did not explain the observation made by Dao (13) on the presence of estrogens in the urine of adrenalectomized oophorectomized women with breast cancer. These early results led some investigators to suggest that the mammary tumor may contribute to steroid synthesis in breast cancer. Since all steroids and their precursors are absent from the system after adrenalectomy and oophorectomy, the only source of steroid hormones would be cholesterol, which is freely available to these tissues.

II. Cholesterol Metabolism

Cholesterol metabolism to steroid hormones is confined to the endocrine glands, adrenal, ovary, testes and placenta. Cleavage of the cholesterol side-chain involves hydroxylation at C-20 and C-22 followed by the action of C-20, C-22 desmolase leading to the formation of pregnenolone. Thus, evidence on C-20, C-22 desmolase in breast tumors is important in delineating the hormonal microenvironment in which the breast tumor continues to grow. Only two published studies have been reported on cholesterol metabolism by mammary tumors. Adams and Wong (14) reported the isolation of ^{14}C-pregnenolone from incubations using 4-^{14}C cholesterol and breast tumor mitochondria; they were also able to isolate ^{14}C-labeled isocaproic aldehyde from incubations of 26-^{14}C cholesterol with breast tumor mitochondria suggesting positive proof for desmolase activity in mammary tumors. Results of Dao et al. (15) also showed the presence of desmolase in mammary tumors. Recent studies in our laboratory using ten primary breast tumors failed to show any desmolase activity. However, in our studies tissue slices were used for incubation and it is possible that dilution of labeled cholesterol by endogenous cholesterol was responsible for our inability to demonstrate C27-desmolase activity.

III. Progestin Metabolism

Little work has been carried out on metabolism of progestins by tumors. Recent studies (16) showed that human mammary tumors can metabolize progesterone to 5α-dihydroprogesterone, 3α-hydroxy-5α-pregnan-20-one, 20α-hydroxy-pregn-4-en-3-one, 20α-hydroxy-5α-pregnan-3-one and 5α-pregnane-3α,20α-diol indicating the presence of 5α-reductase, 3α-hydroxysteroid dehydrogenase and 20α-hydroxysteroid dehydrogenase. In an early study (17), incubations of 17α-hydroxyprogesterone with several mammary tumors and normal mammary tissue showed transformation to androstenedione by only one tumor tissue and one sample of normal mammary tissue indicating the presence of C-17,20-lyase. However, reports by Deshpande and coworkers (18) showed that mammary tumors, both *in vivo* and *in vitro*, synthesized progesterone from pregnenolone indicating the presence of 3β-hydroxysteroid dehydrogenase/5,4-isomerase but not of C-17,20-lyase. These workers suggested that human breast cancers can control intracellular hormone levels but are incapable of initiating steroid hormone synthesis.

Our studies (Table 1) indicated that all mammary tumors investigated were capable of transforming pregnenolone to progesterone (19) suggesting the presence of 3β-hydroxysteroid dehydrogenase/5,4-isomerase activity in these tumors. A number of reduction products were detected in most of our incubations as determined by Lloyd (16). However, none of these tumors were capable of transforming pregnenolone to 17α-hydroxyprogesterone. All eight estrogen receptor negative tumors and 4 out of 8 estrogen receptor positive tumors, with receptor amount of less than 45 fmol/mg protein, converted pregnenolone to 17α-hydroxypregnenolone. Of the eight estrogen receptor negative tumors which showed the presence of 17α-hydroxylase activity, only 5 were capable of cleaving the pregnenolone side-chain while none of the estrogen receptor positive tumors showed the presence of the 17,20-lyase as evidenced by the absence of formation of androstenedione and dehydroepiandrosterone. Furthermore, only 2 out of 8 estrogen receptor negative tumors were capable of synthesizing estradiol. Thus, our results clearly indicate the presence of 17α-hydroxylase and 17,20-lyase in some human mammary tumors. The relationship of receptor content to enzyme activity will be discussed below. Fig. 1 shows the pathway of metabolism of pregnenolone in human mammary tumors.

IV. Androgen Metabolism

The benefit derived from oophorectomy in the treatment of metastatic breast cancer in premenopausal women is primarily due to removal of circulating estrogens. However, one cannot explain the effect of adrenalectomy in postmenopausal patients on a similar basis since it is known that the levels of circulating

Table 1. Metabolism of pregnenolone by human breast carcinoma tissue. Results are expressed as % of total radioactivity recovered.

Subject No.	ER* (fmol/mg)	% Metabolism	% Conversion to					
			P	17OH-P	17OH-Pr	DHEA	4A-dione	E_2
1	0	13.84	2.31	0	2.02	0.37	0.29	0
2	0	10.71	0.53	0	2.25	0.15	0.63	0
3	0	17.86	3.65	0	0.96	0	0	0
4	0	21.84	4.65	0	1.88	0	0	0
5	0	53.22	0.33	0	11.26	2.83	2.71	0.09
6	0	73.86	0.79	0	21.63	4.68	11.31	0.23
7	1	22.41	1.57	0	3.71	0.85	1.36	0
8	3	11.73	3.61	0	1.26	0	0	0
9	11	12.48	8.64	0	0.82	0	0	0
10	17	16.22	6.51	0	1.09	0	0	0
11	22	15.71	4.08	0	1.36	0	0	0
12	45	8.21	4.73	0	0.37	0	0	0
13	48	17.86	7.46	0	0	0	0	0
14	63	11.84	5.92	0	0	0	0	0
15	86	10.71	3.79	0	0	0	0	0
16	162	3.80	2.16	0	0	0	0	0

*Abbreviations used are: P = progesterone; 17OH-P = 17-hydroxyprogesterone; 17OH-Pr = 17-hydroxypregnenolone; DHEA = dehydroepiandrosterone; 4A-dione = 4-androstenedione; E_2 = estradiol; ER = estrogen receptor.

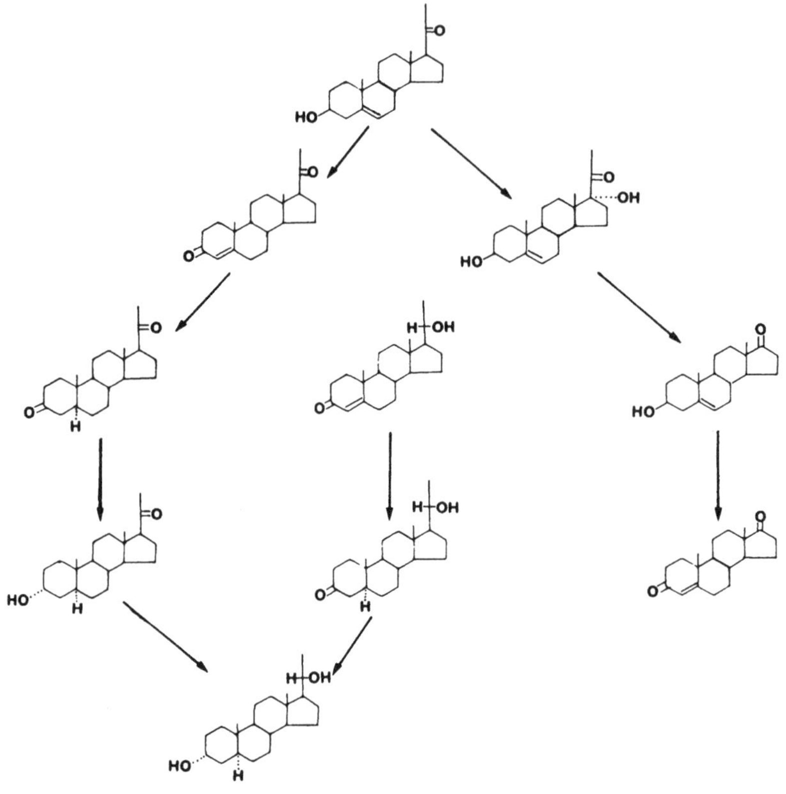

Figure 1. Metabolic pathway of pregnenolone in human breast cancer

estrogens are extremely low (20) and that the adrenal gland secretes only small amounts of estrogens (21). These facts led some investigators to suggest peripheral tissues, including mammary tumors, as sites of estrogen synthesis.

Recently, much attention has been paid to the metabolites of dehydroepiandrosterone (DHEA) as possible mediators in the regulation of mammary tumor growth. High levels of circulating DHEA-sulfate (22) provide a high flux of free DHEA to mammary tumors due to the presence of an active steroid sulfatase (23) which is capable of converting DHEA-S to DHEA. Maynard et al. (24) showed that the levels of free DHEA were considerably greater in mammary tumors than would be expected from contamination by plasma. Thus, the presence of high levels of DHEA would provide sufficient amounts of substrate for metabolism to other physiologically important regulators of tumor growth.

Studies carried out in our laboratory (25) showed the presence of 3β-hydroxysteroid dehydrogenase/5,4-isomerase in human mammary tumors. These studies supported earlier results (17, 23). Even though Li et al. (26) were unable to detect the presence of the isomerase-dehydrogenase in mammary tumors, more recent studies by these investigators showed that mammary tumors were capable of transforming DHEA to Δ^4-3-ketosteroids, indicating the presence of the isomerase-dehydrogenase enzyme system (27). Thus, the isomerase-dehydrogenase system is present in mammary tumors as shown in several other studies (18, 19, 20). Of significant importance is the pathway leading to the formation of 5-androstene-3β,17β-diol (26) which has been shown to cause translocation of cytosol estrogen receptor in both rat uterus and rat mammary tumors (29, 30). Translocation of estrogen receptor in MCF-7 cell cultures by 5-androstene-3β,17β-diol was accompanied by stimulation of protein synthesis, including the specific uterine protein normally induced by estrogens (31). Another important pathway of DHEA metabolism in human mammary tumors is the formation of 7α-hydroxydehydroepiandrosterone. This is believed to be due to direct hydroxylation of DHEA by 7α-hydroxylase (26). 7α-Hydroxylation of 5-androstene-3β,17β-diol results in decreased binding to estrogen receptors which correlates well with the decreased inhibition of estrogen sulfotransferase as a result of 7α-hydroxylation of DHEA. Thus, the high levels of 7α-hydroxylase present in human mammary tumors may function to control intracellular levels of 5-androstenediol and dehydroepiandrosterone.

Even though DHEA-DHEAS is the major circulating androgen in postmenopausal women, most of the studies carried out so far on androgen metabolism by mammary tumors used testosterone as the substrate. Most of these studies (27, 28, 32-42) were carried out after 1974 and were aimed primarily at assessing the ability of mammary tumors to synthesize estrogens. The enzymes, Δ^4-5α-reductase and 3-hydroxysteroid dehydrogenase, were found to be present in the majority of mammary tumors investigated resulting in the formation of 5α-dihydrotestosterone and 5α-androstanediol. Fig. 2 shows the pathway for androgen metabolism in human breast carcinoma. Even though it was believed that androgen metabolites are inactive, these metabolites together with 5-androstenediol may be of significant importance in altering estrogen metabolism. In view of the fact that androgen receptors have been demonstrated in mammary tumors (43, 44), it is possible that androgens may exert either a direct effect on the genesis of breast cancer or they may act indirectly by altering the binding or metabolism of estrogens. Alternatively, androgens may be important in the regulation of estrogen production since 5α-reduced androgens are potent inhibitors of the aromatase enzyme (45). While Δ^4-5α-reductase is shown to be present in most mammary tumors, most reports show no Δ^4-5β-reductase activity. 5β-Reductase has now been demonstrated in breast tumors (40, 46) and 5β-reduction occurred in tumors with very low levels of estrogen receptor.

The aromatase enzyme system is known to be present in human mammary tumors (14, 15, 17, 47). However, it was not until the mid 1970's that unequivocal proof of estradiol formation by human breast carcinoma was demonstrated (25, 32, 36, 39) and many other reports have since appeared (27, 28, 32-42). Cultured MCF-7 breast cancer cells are also capable of aromatizing testosterone to estradiol (48). Thus, it is now well accepted that mammary tumors contain enzymes capable of carrying out various metabolic processes for the synthesis of physiologically active hormones. However, what needs to be elucidated is the physiological significance of hormone synthesis by mammary tumors, especially, estrogen synthesis.

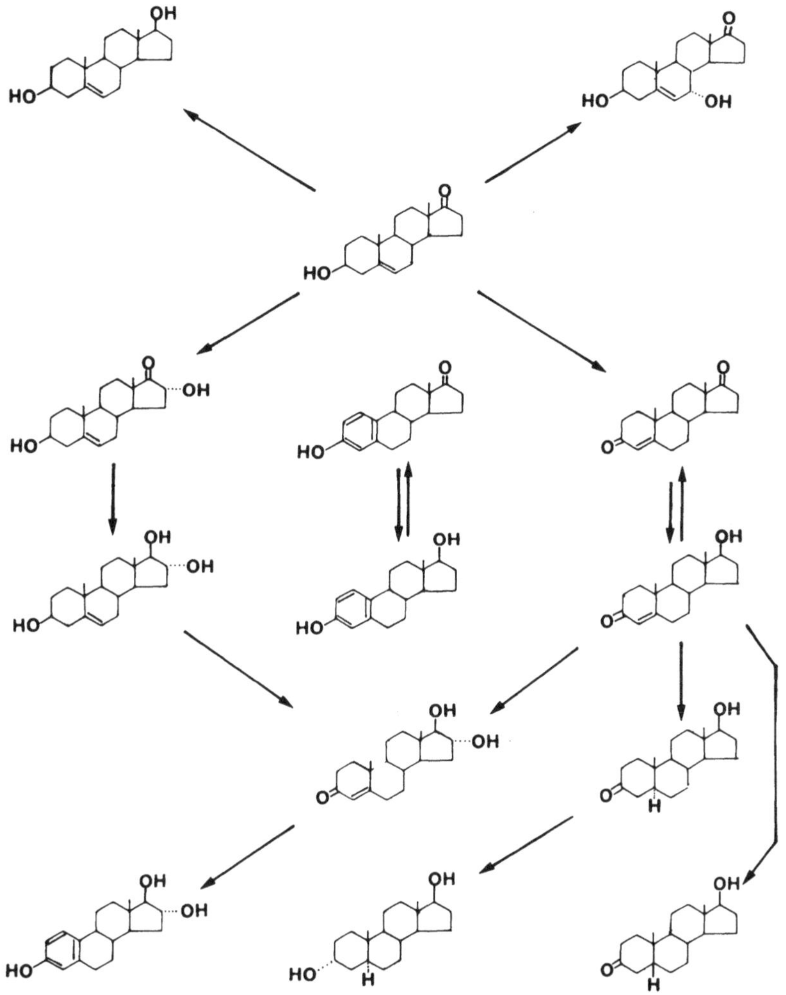

Figure 2. Metabolic pathway of androgens in human breast cancer

Our laboratory has been studying the correlation between estrogen receptor (ER) content and the capacity of mammary tumors to synthesize estrogens. Using DHEA and testosterone, estrogen synthesis occurred predominantly in ER negative tumors

(25, 28) and some ER positive tumors had a lesser biosynthetic capability as evidenced by either the lack of or greatly reduced activities of these enzymes (19, 25, 28, 39). Thus, one possible reason for the lack of response of ER negative tumors to endocrine ablative therapy might be that these tumors were capable of *de novo* synthesis of estradiol. Whether the estradiol produced in ER negative tumors acts by stimulating tumor growth by a mechanism independent of estrogen receptors needs further elucidation. Studies reported by Li and Adams (27) showed that aromatase activity was not confined to ER negative tumors. Even though Varela and Dao (39) found no correlation between the aromatization of androstenedione and the presence or absence of estrogen receptors, analysis of their data show that 8 ER negative tumors (8S) synthesized estrogens at a median of 1.40 pmol/g tumor, or 2.76 $\pm$ 1.57 (mean $\pm$ S.E.M.) whereas 9 ER positive tumors synthesized less estrogens at a median of 0.4 pmol/g tumor, or 1.10 $\pm$ 0.55 (mean $\pm$ S.E.M.). Miller et al. (34, 35) showed that ER positive tumors had a higher capacity for synthesizing estrogens than ER negative tumors and suggested that synthesis of estrogens by ER negative tumors may conceivably be a mechanism by which a hormone-dependent (ER positive) tumor continues to grow in the absence of the endocrine glands. Thus, results reported by several groups of investigators do not agree regarding correlation between receptor content and estrogen synthesis (Fig. 3). Moreover, Bradlow (105) suggested that estrogen synthesis by mammary tumors is of no significant consequence for growth and maintenance of tumors.

Recent studies in our laboratory using GR mouse mammary tumors (49) showed that only hormone dependent tumors were capable of transforming testosterone to estradiol, although both hormone dependent and hormone independent tumors contain aromatase. While the reasons for the inability of hormone dependent tumors to synthesize estradiol are not understood, one possible reason might be the high 5α-reductase activity present in these tumors which may deplete the substrate resulting in decreased aromatization of testosterone.

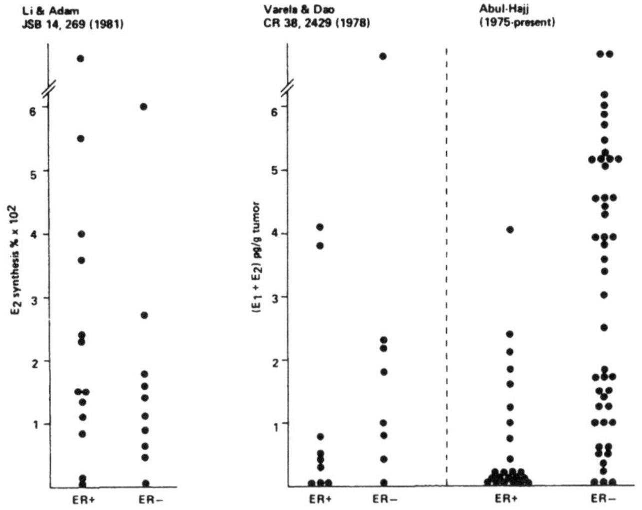

Figure 3 . Relationship between estrogen synthesis and estrogen receptors in human breast cancer

In order to delineate any differences between ER negative and ER positive tumor aromatase, we carried out studies with aminoglutethimide, 4-hydroxy-4-androstenedione and 7α-(4'-amino) phenylthioandrostenedione, which are known to be effective inhibitors of human placental aromatase (106, 107, 108). All three compounds were found to be equally effective inhibitors of tumor aromatase (109, 110), to act by competitive inhibition and the kinetic parameters (K_m, V_{max} and K_i) were similar for both ER negative and ER positive tumors. These studies suggest that aromatase derived from both ER positive and ER negative tumors is essentially the same and may be similar to placental aromatase. In view of the fact that aromatase inhibitors have been shown to effectively block the enzyme in ER negative mammary tumors, experiments to study the effect of aromatase inhibitors on the growth of hormone-independent human mammary tumors may help to explain the significance of estrogen synthesis in ER negative tumors. Recent studies (111, 112) showed that tumors previously responsive to hormonal manipulation were also responsive to aminoglutethimide therapy and that no response was observed in patients with ER negative mammary tumors. These results suggest that estrogen synthesis by ER negative tumors may be unimportant for the growth and maintenance of mammary tumors and therefore are contrary to our hypothesis. However, it is possible that there may be significant differences in the rate of metabolism and distribution of aminoglutethimide between ER negative and ER positive tumors. The growth of mammary tumor cells lacking estrogen receptors required the presence of estradiol in the culture medium (113), suggesting a mechanism for estrogen action in the absence of estrogen receptors. These results enforce our hypothesis on the role of estrogen synthesis in ER negative tumors.

V. Estrogen Metabolism

The role of estrogens in the induction of malignant mammary tumors in laboratory animals is now well established, although there is no evidence to suggest that the same compounds are carcinogenic in man. However, several epidemiological studies implicate estrogens as possible etiologic factors in breast cancer. Most of these studies are based on urinary or plasma levels of various estrogen metabolites and the reader is referred to reviews by Dao (50) and Vorherr (51).

Some studies of *in vitro* metabolism of estrogens by mammary tumor tissues have been reported. James et al. (52) who infused estradiol into breast cancer patients identified in the neoplastic tissue 3-50% of extractable steroids as estrone. Dao et al. (15) demonstrated conversion of estradiol to estrone, 16α-hydroxyestrone and 2-methoxyestrone. The metabolism of estradiol by 31 tumors showed estrone to be the major metabolite and no estriol was detected (53). These results agree with other studies (54, 55, 56, 57). Pollow et al. (56) found that

17β-hydroxysteroid dehydrogenase activity was lower in all malignant mammary tissues compared to nonmalignant tissue samples (including fibroadenomas and cystic mastitis) which may indicate that the "defense mechanism" against the biologically active estradiol becomes less effective in mammary tumors.

Recent studies have shown that the levels of available cytoplasmic estrogen receptors are much lower in premenopausal women than in postmenopausal women (58-61). Variations in receptor concentrations may be related to a partial or total occupancy of receptors by high circulating endogenous estrogens in the premenopausal patient which may lead to false negative results (62, 63). However, several studies (64-67) showed little or no correlation between plasma estrogens and estrogen receptors in breast cancer. Furthermore, even though Sahadevan et al. (68) and Fishman et al. (69) showed ER positive tumors to contain higher levels of estrogens in tumor cytosols, these results do not support the proposal that high levels of endogenous estrogens may be responsible for low receptor measurements. In our study (53), ER negative tumors had a higher 17β-hydroxysteroid dehydrogenase activity than ER positive tumors, which suggested that the observed variations in receptor concentration may be related to variations in 17β-hydroxysteroid dehydrogenase. Estrone levels were higher in ER negative tumor tissues which correlated well with higher 17β-hydroxysteroid dehydrogenase in these tumors (70). Wilking et al. (71) showed that the high rate of estradiol formation from estrone sulfate by mammary tumors may lead to high intracellular levels of this steroid resulting in blockage of the hormone binding site, increased nuclear translocation of the receptor and low cytosol estrogen receptor values.

One of the major pathways of estradiol metabolism involves hydroxylation at C-2 or C-4 and C-16 leading to the formation of catechol estrogens and estriol, respectively. These alternate pathways have attracted considerable interest over the past decade which led Martucci and Fishman (72) to suggest that the direction of metabolism of estradiol may have a profound impact on the nature of the biological response to the hormone. Of significant importance is the relationship between these metabolites and breast cancer. One postulated reason for the lower incidence of breast cancer in Japan, compared to most Western countries (73, 74) is a difference in estrogen levels between the two populations. In support of this hypothesis, several workers have shown that the "estriol ratio" [estriol/(estrone + estradiol)] is higher in Japanese women and other low-risk populations (75, 77) than in Caucasians (75, 76). However, this hypothesis is no longer valid in explaining altered breast cancer risk in certain women or in defined populations based on several lines of evidence including the carcinogenicity of estriol in mice (78), estriol's action as an estradiol agonist, especially when administered chronically (79) and production rates of estriol in women with previous breast cancer were not significantly different from normal women (80). Conversely, there is evidence that the catechol estrogens, 2-hydroxyestrone and 2-hydroxyestradiol, may play the antiestrogen role ascribed to estriol. These metabolites compete with estradiol receptors in the hypothalamus and pituitary (81) and may have a role in the control of gonadotropin secretion (82, 83, 84). On the other hand, 2-hydroxyestradiol, which competes for estrogen receptors in rat uterine tissue (85) and in DMBA-induced rat mammary tumors (86) and causes significant nuclear receptor translocation (86), has been shown to have appreciable uterotrophic (72, 86) and antitumor (87) activity.

Little work has been carried out on assessing estrogen hydroxylase activity in mammary tumor tissues. 16α-Hydroxylation of estrone was reported by Dao et al (15) in 1972. Attempts to show 16α-hydroxylation in our laboratory have failed so far although phenolic polar metabolites have been detected. Unlike 16α-hydroxylase, estrogen 2-hydroxylase has been demonstrated by several investigators leading to the formation of catechol estrogens (88). Aromatic hydroxylation of estradiol at C-2 is the major pathway for estrogen metabolism followed by methylation catalyzed by the enzyme catechol-O-methyl-transferase(COMT),

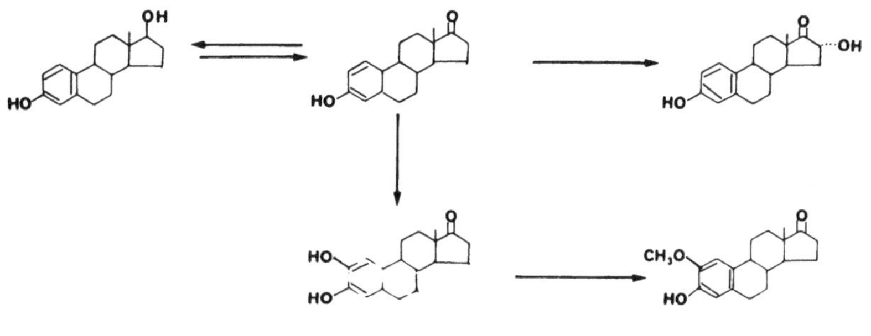

Figure 4. Metabolic pathway of estrogens in human breast cancer

a mechanism analogous to that involved in the metabolism of catecholamines (89). Even though Dao et al. (15) reported 2-methoxyestrone formation from estrone by mammary tumor tissues, unequivocal proof of catechol estrogen was demonstrated only recently by Hoffman et al. (90). Because catechol estrogens have important biological effects (91, 92, 93) and in view of the fact that only some neoplastic tissues had 2-hydroxylase activity while most had greatly increased COMT activity (90, 94) a significant influence of catechol estrogens on the growth and maintenance of hormone-dependent tissues is suggested. The dramatically elevated COMT activity may reflect a detoxifying mechanism whereby catechol-estrogen activity is abolished due to decreased binding to tumor cytosol estrogen receptors. On the other hand, the increase in COMT activity may enhance tumor growth by depleting the pool of anti-estrogenic catechol estrogens (87). Thus, studies on the correlation between receptor content, 2-hydroxylase activity and the effect of catechol estrogens on breast tumor growth will certainly aid our understanding of the role of catechol estrogens in mammary tumors. Fig. 4 shows the pathway for estrogen metabolism in human breast cancer.

VI. Conjugation-Deconjugation

The major circulating androgens and estrogens in postmenopausal women are DHEA-S and estrone-sulfate (E_1S). The levels of these steroids have recently been shown to be even higher in breast cyst fluid (95, 96). Since the free steroids are used for synthesis and metabolism, as well as receptor binding, the presence of sulfatase in mammary tumors is of considerable importance. Mammary tumor homogenates contain appreciable amounts of sulfatase and some of the transformation of steroids by mammary tumors may involve the sulfates forms as intermediates (17).

Another pathway which may be significant in controlling the level of free steroids in tumor tissues is the sulfurylation of DHEA, estradiol and estrone. The presence of alcohol sulfotransferase in human mammary carcinomas was reported in 1964 by Adams (97). Dao and Libby (98) showed that sulfurylation of DHEA and estradiol is of value in predicting the response to endocrine ablative therapy; tumors lacking sulfotransferase activity did not respond to bilateral adrenalectomy while tumors which sulfurylated DHEA more efficiently than estradiol showed the highest remission. Recently Godfroi et al. (99) and Leung et al. (61) have shown a correlation between sulfotransferase activity and estrogen receptors. Adams and Chandra (100) studied the relative sulfation of DHEA and estradiol in 41 tumors and showed that the low levels of DHEA sulfotransferase were responsible for ratios of less than 1. There was a significant correlation between the synthesis of adenosine 3'-phosphate-5'-phosphosulfate (PAPS) and estrogen sulfotransferase activity but not DHEA sulfotransferase activity. More recently a significant correlation was shown between estrogen sulfurylation and ER status (101, 102) and when both ER and progesterone receptors were present in mammary tumors, the levels of estrogen sulfotransferase were significantly higher than in tumors lacking progesterone receptors (103). Thus, these workers have proposed that determination of estrogen sulfotransferase activity in mammary tumors may be a better discriminant than ER for predicting hormonal dependence (102). The levels of steroid sulfotransferase in mammary tumors may be related not only to the hormonal dependence of the tumor tissue but may also be important for regulating the levels of free DHEA and related compounds which may influence the growth of mammary tumors.

Incubation of estradiol with mammary tumor tissue (104) resulted in the formation of a non-polar metabolite called lipoidal estradiol (L_{E_2}). Whilst the structure of L_{E_2} was not conclusively identified, the metabolite may be a fatty acid ester of estradiol at C-17. Estradiol fatty acyl esters are formed in some human mammary tumors and MT/W9a-B rat mammary tumors (114), the oleoyl ester in the non-polar metabolite fraction ranged from 19% to 25% while that of the stearyl acyl ester ranged from 10% to 16%. In all samples the proportion of oleoyl ester in the non-polar fraction was higher than that of the stearate ester. These studies were recently confirmed by Mellon-Nussbaum et al. (115) who showed that the unsaturated fatty acid esters of estradiol comprise more than 85% of the total lipoidal estradiol, estradiol-17β-arachidonate was the most abundant component.

It is interesting to note that the lipoidal estradiol was isolated from tissues and not from incubation medium (104) which may be explained by the fact that these compounds are highly hydrophobic. It was suggested that these compounds may also bind to estrogen receptors in a fashion similar to that described by Korenman (116) for binding of estradiol 17β-acetate to estrogen receptors and that L_{E_2} may play a role in estrogen action by affecting the intracellular equilibrium between estrone and estradiol as well as serve as a long-lived storage form of estradiol. We have also detected the presence of non-polar metabolites in incubations of human breast cancer tissue with dehydroepiandrosterone, testosterone and androsterone.

VII. Concluding Remarks

There is no doubt that human mammary tumors contain several steroid metabolizing enzymes leading to either inactivated products or to biologically active steroid hormones including progesterone, testosterone and estradiol. Mammary tumors metabolize cholesterol to pregnenolone, dehydroepiandrosterone and finally to estrogens, indicating that they have 'para-endocrine' activity. Table 2 summarizes the enzymes that have been detected in human mammary tumors.

Little work has been carried out on the significance of steroid enzymes in mammary tumors; a correlation of aromatase and sulfotransferase with estrogen receptor content has been shown. In the case of aromatase, results from different laboratories are controversial. Studies on sulfotransferase activity are more consistent.

It is interesting to note that mammary tumors behave both as steroidogenic and metabolizing organs. Thus, regulation of the interrelationship between the activities of several enzymes leading to the formation of steroids having a profound effect on mammary tumor development needs further investigation. In particular, the relationship between hormonal dependence and the levels of enzymes leading to the formation of 7α-hydroxy DHEA, catechol estrogens, 5-androstene-3β,17β-diol and estradiol synthesis should be studied.

Table 2. Steroid metabolizing enzymes in human breast cancer

Enzyme	Substrate	Reference
20,22-Lyase	cholesterol	14,15
17α-Hydroxylase	pregnenolone	19
17,20-Lyase	pregnenolone	19
	17α-hydroxyprogesterone	17

Table 2 (continued)

Enzyme	Substrate	Reference
Δ^4-5α-Reductase	progesterone	16,19
	testosterone	28,32,33,36,40
	dehydroepiandrosterone	25,28,40
Δ^4-5β-Reductase	testosterone	40,46
3α-Hydroxysteroid dehydrogenase	progesterone	16
	testosterone	28,32,33,36
	dehydroepiandrosterone	25,28
3β-Hydroxysteroid dehydrogenase	pregnenolone	18,19
	dehydroepiandrosterone	18,25,26,28,29,31
$\Delta^{4,5}$-Isomerase	pregnenolone	18,19
	dehydroepiandrosterone	17,18,23,25,28
17β-Hydroxysteroid dehydrogenase	testosterone	28,32,33,36,40
	androstenedione	39,42
	estradiol	15,52,53,54,55,56,57,70
2-Hydroxylase	estrone	15,90
17α-Hydroxylase	dehydroepiandrosterone	26
16α-Hydroxylase	testosterone	15,17
	dehydroepiandrosterone	17,26
	estrone	16
Aromatase	testosterone	27,28,32,33 34,35,36,37 38,48
	androstenedione	39,41,42
	dehydroepiandrosterone	25,28,47
	pregnenolone	19
Sulfatase	dehydroepiandrosterone	17,23,24,47
Sulfotransferase	dehydroepiandrosterone	26,97,98,100,101,102,103
	estrone	26,97,98,100,101,102,103

Acknowledgements

Work reported from this laboratory was supported in part by grants CA20530 and CA22467 from the National Institutes of Health and two grants from the Graduate School, University of Minnesota.

References

1. Beatson, G.T.: On the treatment of inoperable cases of carcinoma of the mamma: suggestions for a novel method of treatment, with illustrative cases. Lancet 2, 104-107 (1896).

2. West, C.D., Damast, B.L., Sarro, S.D., Pearson, O.H.: Conversion of testosterone to oestrogens in castrated, adrenalectomized human females. J. biol. Chem. 218, 409-418 (1956).

3. Dao, T.L., Morreal, C.E.: Isolation and identification of estriol in urine of castrated and adrenalectomized women receiving 4-^{14}C-testosterone. Steroids 12, 651-666 (1968).

4. Schindler, A.E.: Steroid metabolism of fetal tissues. Conversion of androstenedione to estrone. Am. J. Obstet. Gynecol. 123, 265-271 (1975).

5. Ryan, K.J., Naftolin, F., Reddy, V., Flores, F., Petro, Z.: Estrogen formation in the brain. Am. J. Obstet. Gynecol. 114, 454-460 (1972).

6. Frieden, E.H., Patkin, J.K., Mills, M.: Effect of follicle stimulating hormone upon steroid aromatization *in vitro*. Proc. Soc. exp. Biol Med. 129, 606-609 (1968).

7. Longcope, C., Pratt, J.H., Schneider, S.H., Fineberg, S.W.: Aromatization of androgens by muscle and adipose tissue *in vivo*. J. clin. Endocr. Metab. 46, 146-152 (1978).

8. Bolt, H.M., Göbel, P.: Formation of estrogens from androgens by human subcutaneous adipose tissue *in vitro*. Horm. Metab. Res. 4, 312-313 (1972).

9. Nimrod, A., Ryan, K.J.: Aromatization of androgens by human abdominal and breast fat tissue. J. clin. Endocr. Metab. 40, 367-372 (1975).

10. Rizkallah, T.H., Torell, H.M.M., Kelly, W.C.: Production of estrone and fractional conversion of circulating androstenedione to estrone in women with endometrial carcinoma. J. clin. Endocr. Metab. 40, 1045-1056 (1975).

11. MacDonald, P.C., Rombant, R.P., Siiteri, P.K.: Plasma precursors of estrogen. Conversion of plasma Δ^4-androstenedione to estrone in normal males and nonpregnant normal, castrate and adrenalectomized females. J. clin. Endocr. Metab. 27, 1103-1111 (1967).

12. Siiteri, P.K., MacDonald, P.C.: Role of extraglandular estrogen in human endocrinology. In: "Handbook of Physiological Endocrinology", vol. 2, part 1, Eds. Greep, R.O., Astwood, E., American Physiological Society, Washington, D.C., pp. 615-626 (1973).

13. Dao, T.L.: Method for the separation of four phenolic estrogens in human urine by paper chromatography. Endocrinology 61, 242-255 (1957).

14. Adams, J.B., Wong, M.S.F.: Desmolase activity of normal and malignant human breast tissue. J. Endocr. 44, 69-77 (1969).

15. Dao, T.L., Varela, R., Morreal, C.E.: Metabolic transformation of steroids by human breast cancer. In: "Estrogen Target Tissues and Neoplasia", Ed. Dao, T.L., Chicago University Press, Chicago, pp. 163-179 (1972).

16. Lloyd, R.V.: Studies on the progesterone receptor content and steroid metabolism in normal and pathological human breast tissues. J. clin. Endocr. Metab. 48, 585-593 (1979).

17. Adams, J.B., Wong, M.S.F.: Paraendocrine behavior of human breast carcinoma: *in vitro* transformation of steroids to physiologically active hormones. J. Endocr. 41, 41-52 (1968).

18. Deshpande, N., Carson, P., DiMartino, L., Tarquini, A.: Biogenesis of steroid hormones by human mammary gland *in vivo* and *in vitro*. Europ. J. Cancer 12, 271-276 (1976).

19. Abul-Hajj, Y.J., Iverson, R., Kiang, D.T.: Metabolism of pregnenolone by human breast cancer. Steroids 34, 817-827 (1979).

20. England, P.C., Skinner, L.G., Cottrell, K.M., Sellwood, R.A.: Serum oestradiol in normal women. Brit. J. Cancer 29, 462-469 (1974).

21. Baird, D.T., Uno, A., Melby, J.C.: Adrenal secretion of androgens and estrogens. J. Endocr. 45, 135-136 (1969).

22. Wang, D.Y., Hayward, J.L., Bulbrook, R.D., Kumacka, S., Takatani, O., Abe, O., Lusunomya, U.: Plasma dehydroepiandrosterone and androsterone sulphates, androstenedione and urinary androgen metabolites in normal British and Japanese women. Europ. J. Cancer 12, 951-958 (1976).

23. Jones, D., Cameron, E.H.D., Griffiths, K., Gleave, E.N., Forrest, A.P.M.: Steroid metabolism by human breast tumors. Biochem. J. 116, 919-921 (1970).

24. Maynard, P.V., Bird, M., Basu, P.K., Shields, R., Griffiths, K.: Dehydroepiandrosterone and androstenediol in human pri-

mary breast tumors. Europ. J. Cancer 14, 549-553 (1978).

25. Abul-Hajj, Y.J.: Metabolism of dehydroepiandrosterone by hormone dependent and hormone independent human breast carcinoma. Steroids 26, 488-500 (1975).

26. Li, K., Foo, T., Adams, J.B.: Products of dehydroepiandrosterone metabolism by human mammary tumors and their influence on estradiol receptor binding. Steroids 31, 113-127 (1978).

27. Li, K., Adams, J.B.: Aromatization of testosterone and oestrogen receptor levels in human breast cancer. J. Steroid Biochem. 14, 269-272 (1981).

28. Abul-Hajj, Y.J., Iverson, R., Kiang, D.T.: Aromatization of androgens by human breast cancer. Steroids 33, 205-222 (1979).

29. Adams, J.B., Archibald, L., Clarke, C.: Adrenal dehydroepiandrosterone and human mammary cancer. Cancer Res. 38, 4036-4040 (1978).

30. Poortman, J., Prenen, J.A.C., Schwarz, F., Thijssen, J.H.H.: Interaction of Δ^5-androstene-3β,17β-diol with estradiol and dihydrotestosterone receptors in human myometrium and mammary cancer tissue. J. clin. Endocr. Metab. 40, 373-379 (1975).

31. Adams, J., Garcia, M., Rochefort, H.: Estrogenic effects of physiological concentrations of 5-androstene-3β,17β-diol and its metabolism in MCF-7 human breast cancer cells. Cancer Res. 41, 4720-4726 (1981).

32. Miller, W.R., Forrest, A.P.M.: Oestradiol synthesis by human breast cancer. Lancet ii, 866-868 (1974).

33. Miller, W.R., Shivas, A.A., Forrest, A.P.M.: Factors affecting testosterone metabolism by human breast tissues. Clin. Oncol. 4, 77-85 (1978).

34. Miller, W.R., Hawkins, R.A., Forrest, A.P.M.: Steroid metabolism and oestrogen receptors in human breast cancer. Europ. J. Cancer Clin. Oncol. 17, 913-917 (1981).

35. Miller, W.R., Hawkins, R.A., Forrest, A.P.M.: Significance of aromatase activity in human breast cancer. Cancer Res. 42 (suppl.) 3365S-3368S (1982).

36. Adams, J.B., Li, K.: Biosynthesis of oestradiol in human breast carcinoma tissue and a novel method for its characterization. Brit. J. Cancer 31, 429-433 (1975).

37. Li, K., Chandra, D.P., Foo, T., Adams, J.B., McDonald, D.: Steroid metabolism by human mammary carcinoma. Steroids 28, 561-574 (1976).

38. deThibault de Boesinge, L., Lacroix, E., Eenchante, W., Leusen, I.: Transformation de testosterone en oestrogènes par des coupes de tumeurs mammaires humaines. Annales d'Endocrinologie 35, 691-693 (1974).

39. Varela, R.M., Dao, T.L.: Estrogen synthesis and estradiol binding by human mammary tumors. Cancer Res. 38, 2429 (1978).

40. Abul-Hajj, Y.J.: Steroid Δ^4-5β-reductase in human mammary tumors. Steroids 33, 115-124 (1979).

41. Jenkins, J.A., Ash, S.: Metabolism of testosterone by human breast tumors. Lancet ii, 513-514 (1972).

42. Perel, E., Wilkins, D., Killinger, D.W.: The conversion of androstenedione to estrone, estradiol, and testosterone in breast tissue. J. Steroid Biochem. 13, 89-94 (1980).

43. Wagner, R.K., Jungblut, P.W.: Oestradiol and dihydrotestosterone receptors in normal and neoplastic human mammary tumors. Acta endocr. (Kbh.) 82, 105-120 (1976).

44. Maas, H., Engel, B., Trams, G., Novakowski, H., Stolsebach, G.: Steroid hormone receptors in human breast cancer and their clinical significance. J. Steroid Biochem. 6, 743-749 (1975).

45. Siiteri, P.K., Thompson, E.A. Jr.: Studies of human placental aromatase. J. Steroid Biochem. 6, 317-321 (1975).

46. Raju, U., Sklarew, R.J., Post, J., Levitz, M.: Steroid metabolism in human breast cancer cell lines. Steroids 32, 669-680 (1978).

47. Griffiths, K., Jones, D., Cameron, E.H.D., Gleave, E.N., Forrest, A.P.M.: Transformation of steroids by mammary cancer tissue. In: "Estrogen Target Tissues and Neoplasia", Ed. Dao, T.L., Chicago University Press, Chicago, pp. 151-162 (1972).

48. McIndoe, J.H.: Estradiol formation from testosterone by continuously cultured human breast cancer cells. J. clin. Endocr. Metab. 49, 272-277 (1979).

49. Abul-Hajj, Y.J., Kiang, D.T.: Metabolism of testosterone by GR mouse mammary tumors. Cancer Res. 42, 3510-3513 (1982).

50. Dao, T.L.: Metabolism of estrogens in breast cancer. Biochim. biophys. Acta 560, 397-426 (1979).

51. Vorherr, H.: Sex steroids in relation to breast cancer. In: "Breast Cancer Epidemiology, Endocrinology, Biochemistry and Pathobiology", Urban and Schwarzenberg, Baltimore-Munich, pp. 136-199 (1980).

52. James, F., Braunsberg, H., Irvine, W.T., James, V.H.T.: Radioactivity in human breast tumors after infusion of ^{3}H-oestradiol. Steroids 15, 669-678 (1970).

53. Abul-Hajj, Y.J., Iverson, R., Kiang, D.T.: Estradiol 17β-dehydrogenase and estradiol binding in human mammary tumors. Steroids 33, 477-484 (1979).

54. Melville, E.: Conversion of oestradiol by human breast tumors *in vitro*. Biochem. Soc. Trans. 1, 766-767 (1973).

55. Richardson, S.G., Killen, E.: Metabolism of oestradiol by human mammary tumor supernatants. Cancer letters 2, 299-304 (1977).

56. Pollow, K., Boquoi, E., Baumann, J., Schmidt-Gollwitzer, M., Pollow, B.: Comparison of the *in vitro* conversion of estradiol to estrone of normal and neoplastic human breast tissue. Mol. Cell. Endocr. 6, 333-348 (1977).

57. Geier, A., Horn, H., Levig, J.S., Lichtenstein, E., Finkelstein, M.: Metabolism of ^{3}H-estradiol in human breast cancer in organ culture. Europ. J. Cancer 11, 127-130 (1975).

58. Hahnel, R., Twaddle, E., Vivian, A.B.: Estrogen receptors in human breast cancer. *In vitro* binding of estradiol by benign and malignant tumors. Steroids 18, 681-708 (1971).

59. Wittliff, J.L., Hilf, R., Brooks, W.F., Savlov, E.D., Hall, T.C., Orlando, R.A.: Specific estrogen-binding capacity of the cytoplasmic receptor in normal and neoplastic breast tissues of humans. Cancer Res. 32, 1983-1992 (1972).

60. Spaeren, V., Olsness, S., Brennhovd, J., Efskind, J., Pihl, A.: Content of estrogen receptors in human breast cancers. Europ. J. Cancer 9, 353-357 (1973).

61. Leung, B.S., Fletcher, W.S., Lindell, T.D., Wood, D.C., Krippaehne, W.W., Portland, O.: Predictability of response to endocrine ablation in advanced breast cancer. Arch. Surg. 106, 515-519 (1973).

62. McGuire, W.L., Chamness, G.C., Costlow, H.E., Shepherd, R.E.: Hormone dependence in breast cancer. Metabolism 24, 75-100 (1974).

63. Theve, N.O., Carlström, K., Gustafsson, J.A., Gustafsson, S., Nordenskjöld, B., Sköldefors, H., Wrange, Ö.: Oestrogen receptors and peripheral serum levels of oestradiol in patients with mammary carcinoma. Europ. J. Cancer 14, 1337-1340 (1978).

64. Maas, H., Engel, H., Hochmeister, H., Lehmann, F., Trams, G.: Estrogen receptors in human breast cancer tissue. Am. J. Obstet. Gynecol. 113, 377-382 (1972).

65. Okech, M.O., Orinda, D.A.O., Gatei, D.G., Wasunna, A.O.: Plasma oestradiol and oestrogen receptors in mammary tumors among Africans. East Afr. Med. J. 54, 615-620 (1977).

66. Sakai, F., Saez, S.: Existence of receptors bound to endogenous estradiol in breast cancer of premenopausal and postmenopausal women. Steroids 27, 99-110 (1976).

67. Saez, S., Martin, P.M., Chouvet, C.D.: Estradiol and progesterone receptor levels in human breast adenocarcinoma in relation to plasma estrogen and progesterone levels. Cancer Res. 38, 3468-3473 (1978).

68. Sahadevan, V., Perlia, C.P., Economou, S.G., Sky-Peck, H.: Estrogen and estrogen receptors of breast cancer. J. Surg. Oncol. 7, 467-477 (1975).

69. Fishman, J., Nisselbaum, J., Menendez-Botet, C.J., Schwartz, M.K.: Estrone and estradiol content in human breast tumors: Relationship to estradiol receptors. J. Steroid Biochem. 8, 893-896 (1977).

70. Abul-Hajj, Y.J.: Relationship between estrogen receptors, 17β-hydroxysteroid dehydrogenase and estrogen content in human breast cancer. Steroids 34, 217-225 (1979).

71. Wilking, N., Carlström, K., Gustafsson, S., Sköldefors, H., Tollbom, Ö.: Oestrogen receptors and metabolism of oestrone sulfate in human mammary carcinoma. Europ. J. Cancer 16, 1339-1344 (1980).

72. Martucci, C., Fishman, J.: Direction of estradiol metabolism as a control of its hormonal action-uterotropic activity of estradiol metabolites. Endocrinology 101, 1709-1717 (1977).

73. Doll, R., Muir, C., Waterhouse, J.: Cancer incidence in five continents. II. Springer-Verlag, Berlin (1970).

74. Segi, M., Kurihara, M., Matsuyama, T.: Cancer mortality for steroid sites in 24 countries. No. 5. Sendai. Tohoku University School of Medicine, pp. 1964-1965 (1969).

75. Briggs, M.: Ethnic difference in urinary estrogens. Lancet 1, 324 (1972).

76. MacMahon, B., Cole, P., Brown, J.: Etiology of human breast cancer. J. nat. Cancer Inst. 50, 21-42 (1973).

77. Modan, B., Barell, V., Budin, R., Modan, M.: Dietary factors and cancer in Israel. Cancer Res. 35, 3503-3506 (1975).

78. Rudali, G., Apiou, D., Muel, B.: Mammary cancer produced in mice with estriol. Europ. J. Cancer 4, 39-43 (1975).

79. Clark, J.H., Pasko, Z., Peck, E.: Nuclear binding and retention of the receptor estrogen complex. Endocrinology 100, 91-96 (1977).

80. Pratt, J.H., Longcope, C.: Estriol production rates and breast cancer. J. clin. Endocr. Metab. 46, 44-47 (1978).

81. Davies, I.J., Naftolin, R., Ryan, K.J., Fishman, J., Siu, J.: The affinity of catechol estrogens for estrogen receptors in the pituitary and anterior hypothalamus of the rat. Endocrinology 97, 554-557 (1975).

82. Gethmann, V., Knuppen, R.: Effect of 2-hydroxyestrone on leutropin and follitropin secretion in the ovariectomized primed rat. Hoppe-Seylers Z. physiol. Chem. 357, 1011-1013 (1976).

83. Naftolin, T., Morishata, H., Davies, I.J., Todd, R., Ryan, K.H., Fishman, J.: 2-Hydroxyestrone induced rise in serum LH in the immature male rat. Biochem. biophys. Res. Commun. 64, 905-910 (1975).

84. Paul, S.M., Skolnick, P.: Catechol oestrogens inhibit oestrogen elicited accumulation of hypothalamic cyclic AMP suggesting role as endogenous anti-oestrogens. Nature (Lond.) 266, 559-560 (1970).

85. Martucci, C., Fishman, J.: Uterine estrogen receptor binding of catechol estrogens and of estetrol (1,3,5(10)-estratriene-3,15α,16α,17β-tetrol). Steroids 37, 325-333 (1976).

86. Abul-Hajj, Y.J.: Binding of catechol estrogens to the estrogen receptor of DMBA-induced rat mammary tumors. J. Steroid Biochem. 13, 83-88 (1980).

87. Abul-Hajj, Y.J.: Effect of catechol estrogens on rat mammary tumors. Cancer Res. 39, 4882-4886 (1979).

88. Fishman, J.: Role of 2-hydroxyestrone in estrogen metabolism. J. clin. Endocr. Metab. 23, 207-210 (1963).

89. Axelrod, J., Tomchick, R.: Enzymatic methylation of epinephrine and other catechols. J. biol. Chem. 233, 702-705 (1958).

90. Hoffman, A.R., Paul, S.M., Axelrod, J.: Catechol-estrogen synthesis and metabolism by human breast tumors *in vitro*. Cancer Res. 39, 4584-4587 (1979).

91. Ball, P., Knuppen, R.: Catechol estrogens. Acta endocr. (Kbh.) 93 (suppl.) 232, 1-127 (1980).

92. Paul, S.M., Hoffman, A.R., Axelrod, J.: Catechol estrogens: synthesis and metabolism in brain and other endocrine tissues. In: "Frontiers in Neuroendocrinology", Eds. Martini, L., Ganong, W.F., Raven Press, New York, vol. 6, pp. 203-217 (1980).

93. MacLusky, N.J., Naftolin, F., Krey, L.C., Franks, S.: The catechol estrogens. J. Steroid Biochem. 15, 111-124 (1981).

94. Assicot, M., Contesso, G., Bohoun, C.: Catechol-O-methyltransferase in human breast cancers. Europ. J. Cancer 13, 961-966 (1977).

95. Raju, U., Ganguly, M., Levitz, M.: Estriol conjugates in human breast cyst fluid and in serum of premenopausal women. J. clin. Endocr. Metab. 45, 429-434 (1977).

96. Miller, W.R., Humeniuk, V., Kelly, R.W.: Dehydroepiandrosterone sulfate in breast secretions. J. Steroid Biochem. 13, 145-151 (1980).

97. Adams, J.B.: Enzymic synthesis of steroid sulfates. II. Presence of steroid sulfokinase in human mammary carcinoma extracts. J. clin. Endocr. Metab. 24, 988-996 (1964).

98. Dao, T.L., Libby, P.R.: Steroid sulfate formation in human breast tumors and hormone dependency. In: "Estrogen Target Tissues and Neoplasia", Ed. Dao, T.L., Chicago University Press, Chicago, pp. 181-200 (1972).

99. Godfroi, V.C., Locke, E.R., Singh, D.V., Brooks, S.C.: The steroid alcohol and estrogen sulfotransferases in rodent and human mammary tumors. Cancer Res. 35, 1791-1798 (1975).

100. Adams, J.B., Chandra, D.P.: Dehydroepiandrosterone sulfotransferase as a possible shunt for the control of steroid metabolism in human mammary carcinoma. Cancer Res. 37, 278-284 (1977).

101. Adams, J.B., Pewnin, T., Chandra, D.P., Archibald, L., Foo, M.S.: A correlation between estrogen sulfotransferase levels and estrogen receptor status in human primary breast carcinoma. Cancer Res. 39, 5124-5126 (1979).

102. Pewnin, T., Adams, J.B., Ho, K.P.: Estrogen sulfurylation as an alternative indicator of hormone dependence in human breast cancer. Steroids 39, 47-52 (1982).

103. Pewnin, T., Adams, J.B., Ho, K.P.: A relationship between estrogen sulfotransferase and estrogen and progesterone receptor status in human mammary carcinoma. Cancer Res. 40, 1360-1362 (1980).

104. Schatz, F., Hochberg, R.B.: Lipoidal derivatives of estradiol. Endocrinology 109, 697-703 (1981).

105. Bradlow, H.L.: A reassessment of the role of breast tumor aromatization. Cancer Res. 42, 3382S-3386S (1982).

106. Thompson, E.A., Jr., Siiteri, P.K.: The involvement of human placental microsomal cytochrome P-450 in aromatization. J. biol. Chem. 249, 5373-5378 (1974).

107. Brodie, A.M.H., Schwarzel, W.C., Shaikh, A.A., Brodie, H. J.: The effect of an aromatase inhibitor, 4-hydroxy-androstene-3,17-dione, on estrogen-dependent processes in reproduction and breast cancer. Endocrinology 100,1684-1695 (1977).

108. Brueggemeier, R.W., Floyd, E.F., Counsell, R.E.: Synthesis and biochemical evaluation of inhibitors of estrogen biosynthesis. J. Med. Chem. 21, 1007-1011 (1978).

109. Abul-Hajj, Y.J.: Inhibition of androgen aromatization in human breast cancer. J. Steroid Biochem. 13,1395-1400(1980).

110. Abul-Hajj, Y.J.: Comparative studies of aromatase inhibitors in relation to the significance of estrogen synthesis in human mammary tumors. Cancer Res. (suppl.) 42, 3373S-3377S (1982).

111. Lawrence, B.V., Lipton, A., Harvey, H.A., Santen, R.J., Wells, S.A., Cox, C.E., White, D.S., Smart, E.K.: Influence of estrogen receptor status on response of metastatic

breast cancer to aminoglutethimide therapy. Cancer 45, 786-791 (1980).

112. Asbury, R.F., Bakemeier, R.F., Folsch, E., McCune, C.S., Savlov, E., Bennett, J.M.: Treatment of metastatic breast cancer with aminoglutethimide. Cancer 47, 1954-1958 (1981).

113. Dao, T.L., Sinha, D.K., Nemoto, T., Patel, J.: Effect of estrogen and progesterone on cellular replication of human breast tumors. Cancer Res. 42, 359-362 (1982).

114. Abul-Hajj, Y.J.: Formation of estradiol 17β-fatty acyl 17-esters in mammary tumors. Steroids 40, 149-156 (1982).

115. Mellon-Nussbaum, S.H., Ponticorvo, L., Schatz, F., Hochberg, R.B.: Estradiol fatty acid esters. J. biol. Chem. 257, 5678-5684 (1982).

116. Korenman, S.G.: Comparative binding affinity of estrogens and its relation to estrogenic potency. Steroids 13, 163-169 (1969).

BIOCHEMICAL PROPERTIES OF STEROID METABOLIZING ENZYMES

W. Gibb and R. Ducharme
Paediatric Research Centre, Hôpital Sainte-Justine and Departments of Obstetrics and Gynaecology and Paediatrics, University of Montreal, 3175 Chemin Sainte-Catherine, Montreal, Quebec H3T 1C5, Canada.

Introduction

This chapter presents a description of the assays for the different enzymes, their reaction mechanisms, distribution and composition, which in many cases is multimetric. Many of the steroid-converting enzymes have as their ultimate catalytic unit some form of cytochrome P-450, which provides the active-site for various hydroxylation and carbon-carbon cleavage reactions and a brief account of the properties of the drug hydroxylating cytochrome P-450 is provided.

Cytochrome P-450

Cytochrome P-450, a mixed function oxidase, is the terminal oxidizing component in the hydroxylation and C-C bound cleavage of steroids and other lipophylic molecules. It was originally described in mammalian liver for drug hydroxylations (1, 2) and its name results from its ability to form a complex with carbon monoxide which absorbs light at 450 nM (3). The reaction catalyzed by cytochrome P-450 is

$$RH + O_2 + NAD(P)H + H^+ \longrightarrow ROH \longrightarrow NAD(P)^+ + H_2O$$

where RH represents substrate and ROH product.

Steroid Converting Enzymes and Diseases

It is found in the membranous components of cells, mitochondria and endoplasmic reticulum and as it can only be solubilized by detergents, it would appear to be an integral component of the membrane. In microsomes the transfer of electrons from NADPH to the cytochrome is catalyzed by a membrane bound flavoprotein, NADPH-cytochrome P-450 reductase and in mitochondria and bacteria an iron-sulfur containing protein is also involved in this transfer (4).

Studies on the interaction between the substrate and the cytochrome can be carried out by difference spectroscopy which allows the binding constants of various substances to be obtained. This spectrum results as the substrate binding disturbs the equilibrium between the high and low spin states of the ferric haem-iron, resulting in a type I difference spectrum (5). The overall mechanism of the cytochrome P-450 catalyzed reaction is shown in Fig. 1 (6).

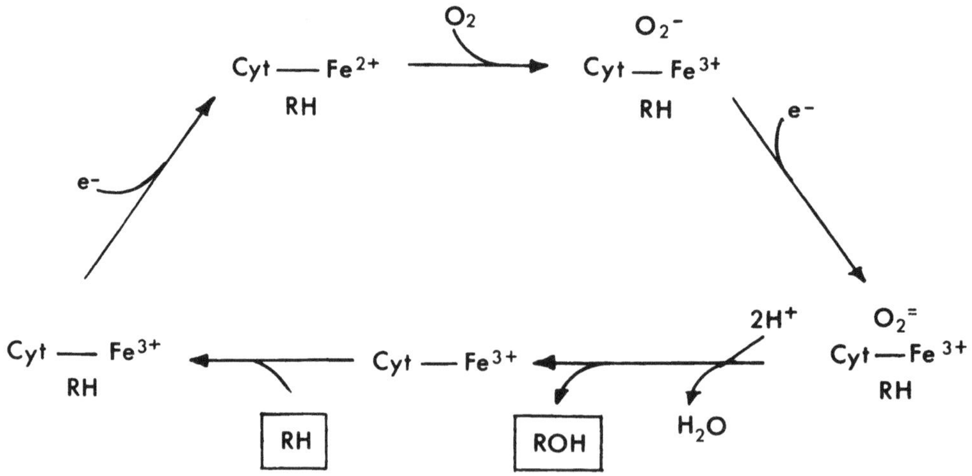

Fig. 1 Reaction scheme for cytochrome P-450 catalyzed hydroxylations. (RH = substrate; ROH = product).

Substrate (RH) binds to the oxidized cytochrome (Fe^{3+}) and interacts with oxygen and two electrons such that the substrate is hydroxylated, a molecule of water is formed and the oxidized form of the cytochrome is regenerated.

Of particular importance in clinical endocrinology is the fact that certain nitrogenous compounds (e.g. metyrapone, aminoglutethimide) interact with cytochrome P-450. The degree of binding of a particular substance depends upon the cytochrome. The various substances which bind to cytochrome P-450 have been reviewed (7). Also, these substances have been used as evidence for the presence of cytochrome in instances where carbon monoxide inhibition of a reaction could not be demonstrated (e.g. 8).

Several excellent articles on various aspects of cytochrome P-450 biochemistry and physiology have appeared (4, 6, 7, 9, 10).

Cholesterol Side-Chain Cleavage

The cholesterol side-chain cleavage reaction (Fig. 2) is catalyzed by a cytochrome P-450 containing enzyme (11, 12) which

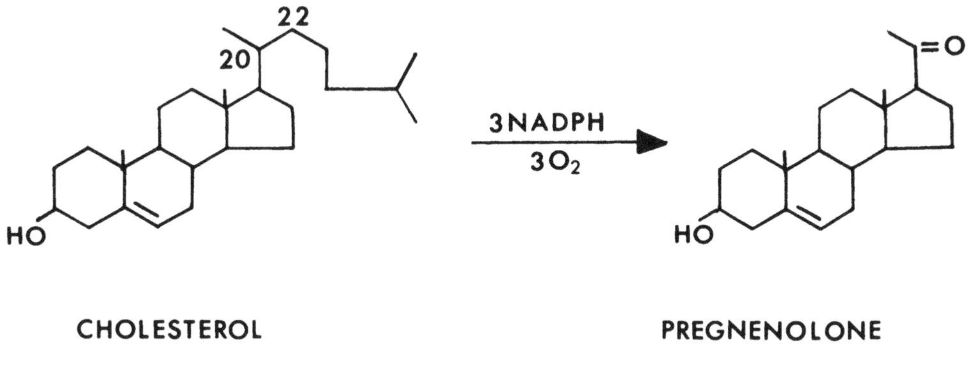

Fig. 2 Reaction catalyzed by cholesterol side-chain cleavage enzyme or desmolase.

converts cholesterol to pregnenolone and is found in a wide variety of steroid producing tissues. The product of the reaction, pregnenolone, is the obligatory precursor of all steroid hormones. This enzyme is considered to be the rate-limiting step in steroid hormone synthesis (12). It is found in the mitochondrial fraction of cells and would appear to be present in the inner membrane (13, 14). In contrast to most other steroid hydroxylations, this enzyme requires two membrane-bound proteins to transfer electrons from the NADPH to the cytochrome, an FAD-containing flavoprotein and an iron-sulphur protein (15, 16). This latter protein in the adrenal is referred to as adrenodoxin and in the testis as testodoxin.

The cytochrome P-450 responsible for the side-chain cleavage reaction (cytochrome $P\text{-}450_{scc}$) has been purified from a number of sources and reconstituted systems studied (17-24). The cytochrome $P\text{-}450_{scc}$ from adrenal mitochondria has been shown to be different from the adrenal mitochondrial cytochrome $P\text{-}450_{11\beta}$ which catalyzes 11β-hydroxylation (25, 26). The amino acid composition of cytochrome $P\text{-}450_{scc}$ has been reported (27) and the cell free synthesis of the cytochrome studied (28, 29). Evidence has been presented for the presence of multiple side-chain cleavage enzymes (30) and the substrate specificity of the reaction has been investigated (31-34).

The cleavage reaction uses three molecules of oxygen and NADPH (35) and is considered to involve two hydroxylations followed by C-C bound cleavage (Fig. 3). The first hydroxylation likely occurs at position 22 and the second at position 20 (36-38). The C-C bond between C-20 and C-22 is then cleaved using the third molecule of NADPH and oxygen and the products pregnenolone and isocaproic aldehyde are formed (39). Using ^{18}O it was shown that the oxygen introduced into position 20 is present in the pregnenolone produced (40). It has been suggested that these hydroxylated steroids are not in fact intermediates but are formed from the transition-states of the enzymatic reaction (41). This suggestion resulted from the fact that an analogue of cholesterol, which could not be hydroxylated

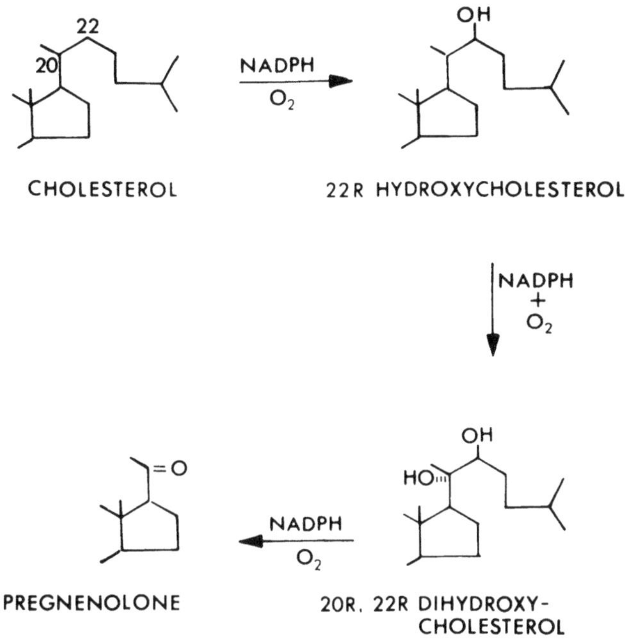

Fig. 3 Possible mechanism for the cholesterol side-chain cleavage reaction. R corresponds to the absolute configuration.

at position 22, was converted to pregnenolone.

A number of assays have been proposed for this reaction. Using radioactive substrate the steroid product can be isolated (22), or [26-^{14}C] cholesterol, the radioactive water soluble side-chain can be separated from the steroid (42, 43). Radioimmunoassay of pregnenolone has been used to assay the enzyme. The reaction is carried out in the presence of an inhibitor of 3β-hydroxysteroid dehydrogenase (3β-HSD), cyanoketone, which prevents transformation of the pregnenolone to progesterone (44). More detailed information on this reaction can be obtained from recent reviews (12, 45).

17α-Hydroxylase/C-17,20-Lyase

The 17α-hydroxylase/C-17,20-lyase reaction, converting C-21 steroids into C-19 ones, is achieved through hydroxylation at position 17 followed by C-C cleavage (Fig. 4). The majority of

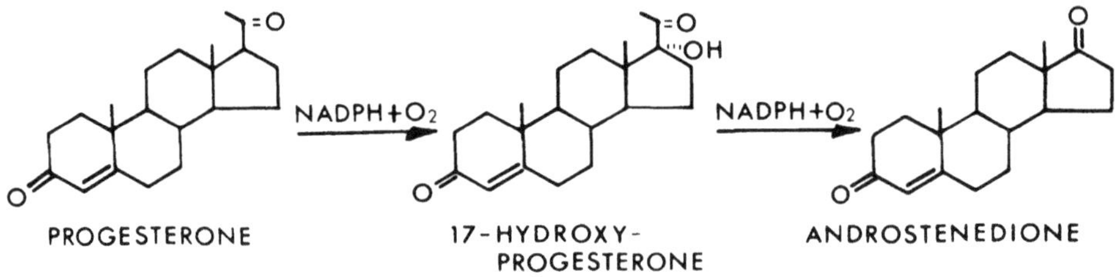

Fig. 4 Reaction catalyzed by 17α-hydroxylase/C-17,20-lyase.

studies on this transformation have been carried out using preparations from rat testis and more recently pig testis, but the reaction has also been studied in other steroidogenic tissues such as the ovary (66), human fetal adrenal (47), and human testis (48-50). In all tissues studied the majority of the activity has been found in the microsomal fraction.

Each reaction uses one molecule of oxygen and one molecule of NADPH. In experiments with ^{18}O it has been shown that the oxygen introduced by the 17-hydroxylase is found at C-17 of the C-19 product (51). The transformations are catalyzed by cytochrome P-450 containing enzyme(s) (52-55) and the transfer of electrons from NADPH to the cytochrome is catalyzed by a flavoprotein reductase (56, 57).

Inhibition studies are consistent with separate cytochromes being responsible for the hydroxylase and the lyase reactions (58). However, it has been suggested (59) that the cleavage reaction is a concerted process and that the 17-hydroxysteroid is not a true intermediate, but is formed from the transition-state of the reaction, a situation somewhat similar to that proposed for the cholesterol side-chain cleavage system. Recently, a cytochrome P-450 from neonatal pig testes has been purified and the hydroxylase and lyase reactions shown to be catalyzed by a single protein. The purified cytochrome is a glycoprotein with a molecular weight of 59,000 daltons (60). The porcine enzyme metabolizes pregnenolone more efficiently than progesterone (61) and is inhibited by oestrogens (62). With the purified enzyme, kinetic studies indicate that the 17-hydroxysteroid is an obligatory intermediate (63).

In rat testis microsomes kinetic analysis of the 17α-hydroxylase has shown that steroid substrates have Km values in the nanomolar range of concentrations (64), compared to previous studies which reported values 1,000-times higher.

A few studies have described the structural features important for steroid binding to these enzymes (47, 65). For example, reduction at C-20 of the 17-hydroxysteroid produces steroids which are no longer substrates for the lyase reaction, but may be potent inhibitors.

These reactions have been assayed with radiolabelled steroids and separating the products by chromatography (49), with a [21-^{14}C] labelled substrate and following the formation of the water-soluble side-chain fragment (66) and by a rapid assay for the 17α-hydroxylase with [17α-^{3}H]-progesterone as substrate (67) in which the tritiated water formed in the reaction is measured.

3β-Hydroxysteroid Dehydrogenase/Δ^5-3-Oxosteroid Isomerase

The formation of all naturally occurring active steroid hormones requires the participation of the 3β-hydroxysteroid dehydrogenase (3β-HSD) and the Δ^5-3-oxosteroid isomerase to transform Δ^5-3β-hydroxysteroids into Δ^4-3-oxosteroids (68, 69). 3β-HSD catalyzes the first reaction (Fig. 5) and is the rate-limiting

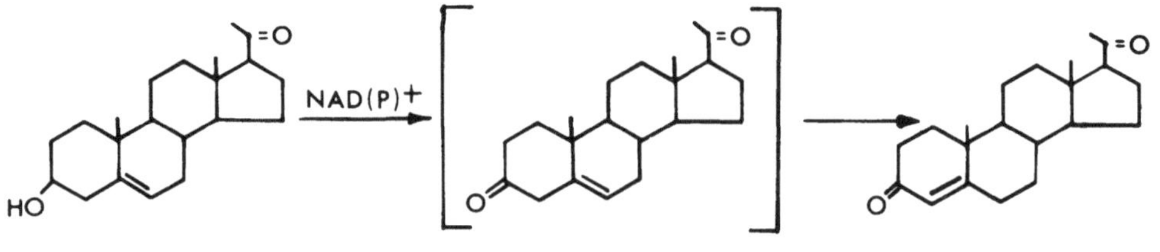

Fig. 5 Reactions catalyzed by the 3β-hydroxysteroid dehydrogenase and Δ^5-3- oxosteroid isomerase.

step in the transformation. The two enzymes are found closely associated in the microsomal and mitochondrial fractions of cells (68-76). Although some evidence indicates that the activity in mitochondria might be due to microsomal contamination (77), other studies have shown that this is unlikely to be the case in all tissues (75, 78, 79).

Both NAD^+ and $NADP^+$ can act as cofactors for 3β-HSD, but NAD^+ is considered to be the preferred cofactor (70, 80-83) and it has been suggested that the activity with $NADP^+$ may be mediated by a transhydrogenase (81). *In vitro*, pregnenolone, 17α-hydroxypregnenolone and dehydroepiandrosterone are capable of serving as substrates for the enzyme. Whether or not the same enzyme metabolizes C-19 and C-21 steroids is open to question. Some studies suggest multiple enzymes whereas others are consistent with a single enzyme metabolizing these steroids (83-87).

Recent studies on the placenta (83, 88, 89) have found that K_m values for steroid substrates are at nanomolar concentrations compared to micromolar concentrations previously found and certain steroids, such as oestrogens, bind to the enzyme with high affinity (83, 87). The presence of sulphydryl protecting agents (β-mercaptoethanol) may be responsible for the low

K_m values (88). NADH produced by 3β-HSD may be involved in the 17-hydroxylase/C-17,20-lyase reaction (90).

A number of steroids inhibit this reaction and may be involved in the control of steroid synthesis in certain tissues (91-94). Solubilized preparations of the enzyme have been examined (95-98) but only in one study does it appear to have been separated from the Δ^5-isomerase (98).

The enzyme can be assayed with radioactive substrates and separating substrate and product by chromatography (89) or by precipitation with digitonin (91). Assays involving u.v. spectroscopy (72) and fluorescence measurements (88) have also been used.

Initial studies of Δ^5-3-oxosteroid isomerase were carried out with an enzyme from Pseudomonas testosteroni and many of the studies of the mammalian enzyme have been based upon information obtained in the bacterial system (99-107). The isomerase is one of the fastest enzyme-catalyzed reactions and would appear not to require a cofactor. Although earlier studies suggested that this reaction might be reversible (108-110), for practical purposes it should be considered irreversible. As for 3β-HSD, the evidence is conflicting as to whether a single or multiple enzymes are responsible for the metabolism of C-19 and C-21 steroids (69, 86, 111-113).

In contrast with the purified bacterial enzymes which have very high turnover numbers, mechanistic studies on the mammalian microsomal enzyme have been complicated by the non-enzymatic isomerization of Δ^5-3-oxosteroids which occurs in aqueous solution. Thus it is tentatively considered that the reaction mechanism is similar to that of the bacterial enzymes, involving removal of the 4β-hydrogen and insertion of hydrogen in the 6β-position (Fig. 6) (114, 115). Also, recent studies have shown that serum albumin markedly increases the activity of the enzyme (116).

This enzyme is usually assayed by following the formation of the Δ^4-3-oxosteroid by way of the increase in absorbtion at 240nM.

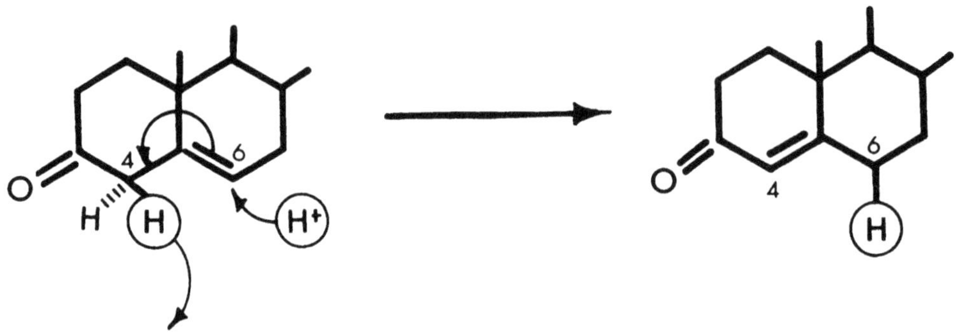

Fig. 6 Possible mechanism for the Δ^5-isomerase.

Aromatase/Oestrogen Synthetase

The aromatase of oestrogen synthetase catalyzes the transformation of C-19 Δ^4-3-oxosteroids (androgens) to C-18 phenolic A-ring oestrogens (Fig. 7).

This enzyme is present in a wide variety of mammalian tissues, ovary, placenta, fetal membranes, testis, brain, adipose tissue, etc. It is generally found in the microsomal fraction (117) but in certain instances it may be present in mitochondria or heavier membrane fractions (118-120). Most studies on aromatase have used the enzyme from human placenta, in contrast with studies on other steroid metabolizing enzymes, which have usually been performed with animal tissues.

Aromatase is a cytochrome P-450 containing enzyme (8), and has been partially purified from placenta (121-123). A flavoprotein is involved in the transfer of electrons from NADPH to cytochrome P-450 (124, 125). For a number of years the participation of cytochrome P-450 in the placental reaction was questioned as carbon monoxide was not found to inhibit aromatization of androstenedione or testosterone. However, carbon

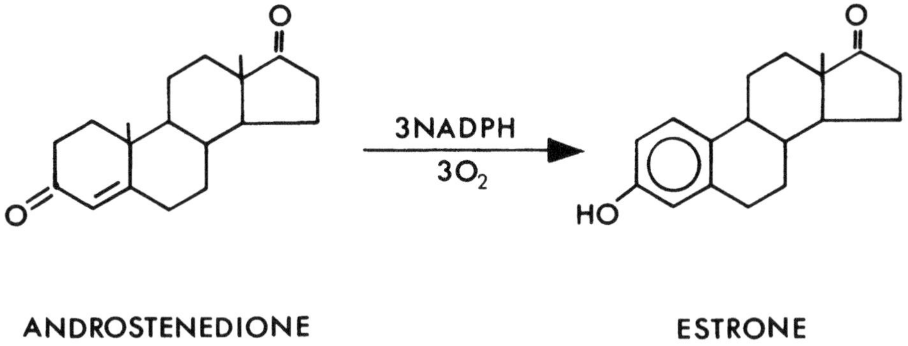

Fig. 7 Reaction catalyzed by aromatase (oestrogen synthetase)

monoxide does inhibit aromatization of 16α-hydroxytestosterone and 19-nortestosterone (126, 127). In contrast with the placental enzyme, ovarian aromatization is sensitive to carbon monoxide inhibition (128).

Multiple cytochrome P-450's have been isolated from placental microsomes (123), indicating the presence of multiple enzymes. Since detailed kinetic studies indicate that a single aromatase metabolizes testosterone and androstenedione (129), it is possible that a separate cytochrome P-450 is responsible for the aromatization of 16-hydroxylated steroids. The reaction uses three molecules of oxygen and NADPH (130). Kinetic studies indicate that a single active-site is responsible for the reaction, with the intermediates of the placental enzyme being tightly bound to the active-site (131). However, a recent study has provided other kinetic evidence that the final hydroxylation (see following paragraph) in fact occurs at an active-site different from that of the first two hydroxylations (132).

The mechanism involved in aromatization is shown in Fig. 8. There are two hydroxylations at C-19 and the gem-diol is then transformed to the 19-aldehyde by removal of water (133-141).

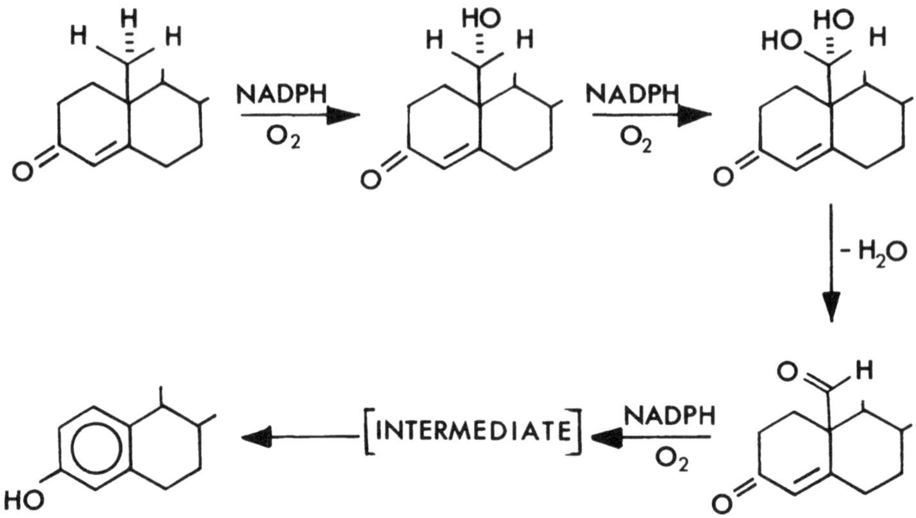

Fig. 8 Mechanism of steroid aromatization

The third hydroxylation and final transformation of the Δ^4-3-oxosteroid to the phenolic oestrogen has been the subject of debate. Evidence has been presented supporting the involvement of 1β-hydroxylation (142), 2β-hydroxylation (143-145) and epoxide formation (146). However, the precise reaction mechanism is still open to further study.

The aromatase has been measured with radioactive substrates (129), and the formation of tritiated water (130, 147). The latter assay results from the finding that the 1β and 2β-hydrogens are removed during aromatization (142, 148, 149) and therefore tritiated water is formed when a substrate labelled with tritium in these positions is used. A number of reviews on aromatization have been published (150-152).

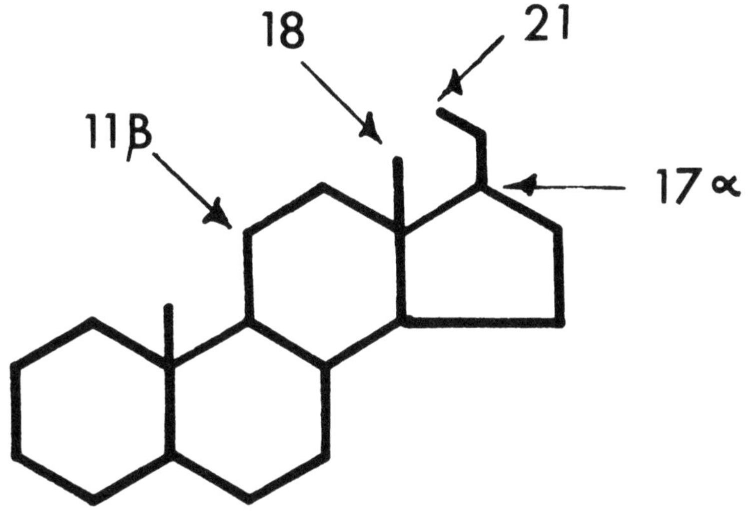

Fig. 9 Positions of the steroid nucleus hydroxylated by adrenal tissue

Steroid Hydroxylases

A large number of steroid hydroxylases exist (Fig. 9). Those most intensively studied are involved in adrenal steroid biosynthesis: 11β-hydroxylase, 17α-hydroxylase, 18-hydroxylase and 21-hydroxylase. All are cytochrome P-450 containing enzymes. Reviews describing these hydroxylases have been published (45, 153).

11β-Hydroxylase, like cholesterol side-chain cleavage enzyme, is found in the inner membrane of mitochondria (13, 14). The cytochrome $P\text{-}450_{11\beta}$ has been purified to homogeneity and shown to be different from cytochrome $P\text{-}450_{scc}$ (25, 26). Two proteins are required to transfer electrons from NADPH to the cytochrome, adrenodoxin and adrenodoxin reductase. These proteins have been purified and characterized and systems, reconstituted with purified cytochrome-P-$450_{11\beta}$, have been studied (154-166). The cell free synthesis of the cytochrome has recently been described (28).

The 11β-hydroxylase can hydroxylate androgens at C-11β and C-19 (45, 167) and the highly purified cytochrome can hydroxylate deoxycorticosterone on both C-11β and C-18 (25, 167-169). However, 18-hydroxylation of corticosterone, a major step in aldosterone biosynthesis, is not catalyzed by the purified cytochrome P-$450_{11\beta}$ (25, 167).

In contrast with 11β-hydroxylation, 21-hydroxylation occurs in the microsomal fraction (170). This reaction was shown to be dependent upon cytochrome P-450 (171, 172) and as it is present in the microsomal fraction and not the mitochondria, only one flavoprotein is involved in the transfer of reducing equivalents from NADPH to the cytochrome (173, 174). The cytochrome P-450 $_{21OH}$ and the NADPH cytochrome P-450 reductase have been purified and reconstituted systems studied (175-178). The cell free synthesis of this cytochrome has recently been described (28). The highly purified cytochrome P-450_{21OH} is a glycoprotein with a molecular weight of 47,000 daltons and is immunologically different from the adrenal cytochromes responsible for the side-chain cleavage of cholesterol and 11β-hydroxylation (176, 177). The bovine adrenal enzyme was shown to be activated by cytosol (179) and recent studies indicate that this activation is due to multiple factors present in the cytosol (180).

Aldosterone is produced by the zona glomerulosa (181) and its synthesis occurs in two stages (182-184). The first is 18-hydroxylation of corticosterone (184, 185). The second step has been proposed to involve an 18-dehydrogenase or possibly a second hydroxylation at C-18 and removal of water to form aldosterone (181, 186). 11β-Hydroxylase can catalyze 18-hydroxylation, but there are apparently different 11β- and 18-hydroxylating enzymes in the zona fasciculata and zona glomerulosa (187). More information on aldosterone biosynthesis can be obtained from a recent review (188).

17-Hydroxylation like 21-hydroxylation takes place in the microsomal fraction of cells (189). With the recent demonstration that the testicular 17-hydroxylase/C-17,20-lyase reaction is catalyzed by a single protein (60), future studies will be required to determine if the adrenal 17-hydroxylase, involved in the synthesis of glucocorticoids, is different. Hydroxylase reactions are usually assayed with radiolabelled substrates and separation of the products by chromatography.

Steroid Oxidoreductases

These enzymes catalyze the reversible oxidation and reduction of steroids, usually hydroxyl and oxo groups respectively. They are widely distributed in mammalian tissues and have been studied from a number of sources. Oxidoreductases are found both in the soluble-protein fraction of cells and in the microsomal fraction. Because of the vast literature on these enzymes, this review will concern mainly the soluble human placental enzyme first reported by Ryan and Engel (190) and named oestradiol dehydrogenase (191).

The reaction catalyzed by the enzyme is shown in Fig. 10.

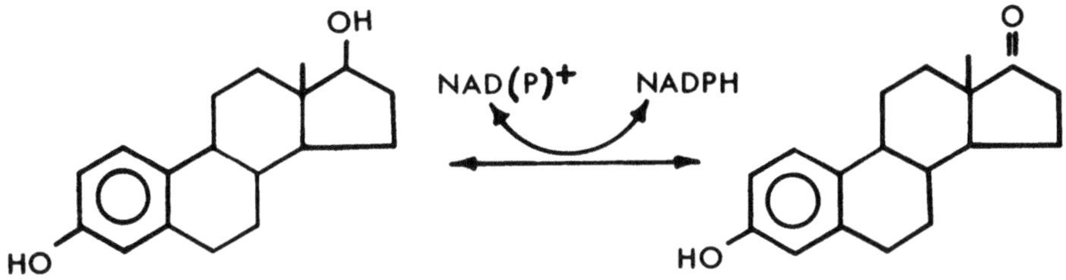

Fig. 10 Reaction catalyzed by oestradiol dehydrogenase

It is a 17β-hydroxysteroid oxidoreductase which can use NAD^+ or $NADP^+$ as cofactor. It also has transhydrogenase activity (192). The enzyme was found to reduce 16α-hydroxyoestrone irreversibly and it was suggested that its physiological function was to catalyze this final step in oestriol synthesis (193).

The purification and crystallization of the enzyme by a number of groups have allowed it to be studied in great detail (194-197). Its amino acid composition has been described (195, 198-200) and a number of laboratories have studied the amino acids associated with the active-site by affinity labelling (e.g. 193, 200-205). With this technique, steroids (or nucleotides), with reactive groups at different positions on the nucleus, are allowed to interact with the enzyme; then after hydrolysis, the amino acids attached to the reactive moiety are isolated. In this manner, by using a number of steroids, with reactive groups at different positions on the nucleus, the amino acid composition of the active-site can be defined. Other investigators have described the active-site by using different steroids and comparing their binding properties (200).

The placental oestradiol dehydrogenase has recently been shown to be responsible for the 20α-hydroxysteroid dehydrogenase activity found in placenta, both 17β- and 20α-activities occurring at the same active-site (206). This situation, whereby a single active-site catalyzes reactions at two positions on the steroid nucleus, occurs with other steroid oxidoreductases (207-210).

Steroid oxidoreductases are not only stereospecific with respect to the hydroxyl group on the steroid, but like all other oxidoreductases transfer hydrogen stereospecifically to either the A or B side of the nicotinamide ring in the nucleotide. For example, oestradiol dehydrogenase transfers hydrogen to and from the B side of the nucleotide (211, 212). In Fig. 11 the situation with a 3β-hydroxysteroid dehydrogenase is shown. The hydrogen atoms on the 4-position of the nicotinamide ring of the reduced nucleotide are not identical being either A or B (pro-R or pro-S respectively when using absolute configurations).

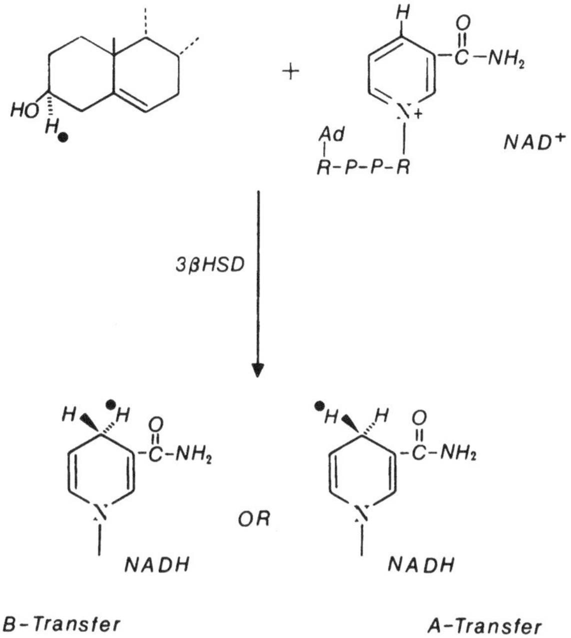

Fig. 11 Hydrogen transfer from steroid to the nicotinamide ring of NAD^+ catalyzed by a dehydrogenase.

As the stereospecificity of hydrogen transfer to the nucleotide is thought always to be the same for each enzyme, this specificity has been used to investigate whether or not a single active-site metabolizes different substrates (72).

More detailed information on oestradiol dehydrogenase is given in a number of reviews (193, 200, 213, 214). Dehydrogenases are usually assayed by following the formation of NAD(P)H at 340nM or by fluorescence.

Steroid Sulphotransferases

Steroid sulphates are not only excretion products but also serve as precursors of active hormones (215). For example, during human pregnancy oestrogens are synthesized from dehydroepiandrosterone sulphate which is produced by both the maternal and foetal adrenal. The enzymes responsible for sulphating steroids are sulphotransferases (Fig. 12). They are found in the soluble-protein fraction of the cell and require PAPS (3'-phospho-adenosine-5'-phosphosulfate) for activity, this compound provides "activated sulphate" synthesized from ATP (215). Sulphotransferases have been studied from adrenal, testis, ovaries and liver. There are specific enzymes to sulphate oestrogens, glucocorticoids and Δ^5-3β-hydroxysteroids.

An oestrogen sulphotransferase has been purified from bovine adrenals and found to have complex kinetics (216). This enzyme is different from the adrenal dehydroepiandrosterone sulphotransferase (217) but identical to the oestrogen sulphotransferase in bovine placenta (218). Structural analogues of PAPS which can act as co-substrates and inhibitors of this enzyme have been studied (219, 220).

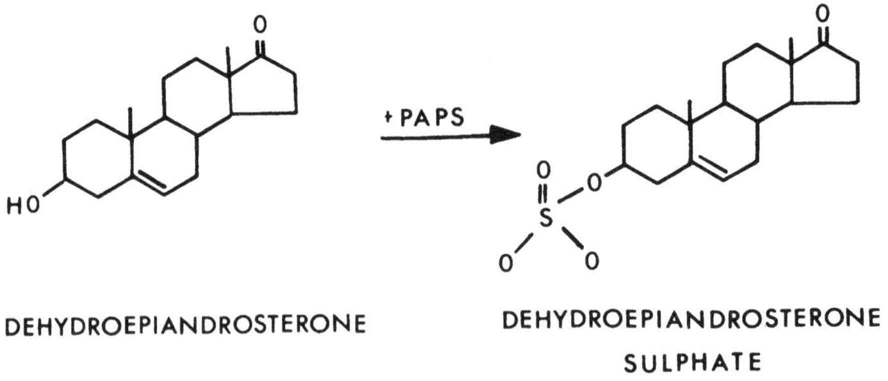

Fig. 12 Reaction catalyzed by steroid sulphotransferase

The 3β-hydroxysteroid sulphotransferase from rat liver which sulphates dehydroepiandrosterone and pregnenolone but not oestrogens (221) has been purified to homogeneity (222). This purified enzyme has a molecular weight of 180,000 daltons and is composed of subunits of 28,000 daltons (222). Recently, the human adrenal enzyme, responsible for sulphating dehydroepiandrosterone, was purified from both adult and foetal tissues (223, 224). The enzyme from both tissues are identical. The purified enzyme has a molecular weight of 68,000 daltons, is composed of two subunits and like other sulphotransferases, exhibits complex enzyme kinetics. The 3β-hydroxysteroid sulphotransferase has also been studied in testicular tissues (225, 226).

One of the most carefully studied sulphotransferase systems is that responsible for the sulphation of glucocorticoids by rat liver. Three separate glucocorticoid sulphotransferases have been described (227) and the endocrine factors which control their production have been examined (228-232). The properties of the partially purified enzymes have also been studied (233-236).

The assay systems for sulphotransferases usually involve radioactive steroid substrates or [^{35}S]PAPS. A recent study (237) compared two methods: one using radioactive steroid substrate and separation of substrate and product by filter binding, and the other using [^{35}S]PAPS and isolating the radioactive product by thin-layer chromatography. For more information, the reader is referred to reviews (215, 238).

Steroid Sulphatase

Steroid sulphatases are widely distributed in mammalian tissues and are responsible for the hydrolysis of sulphated steroids (215) (Fig. 13). The enzymes are microsomal and do not require cofactors. Only recently have solubilized, purified preparations of the enzymes been obtained from liver and placenta. In the

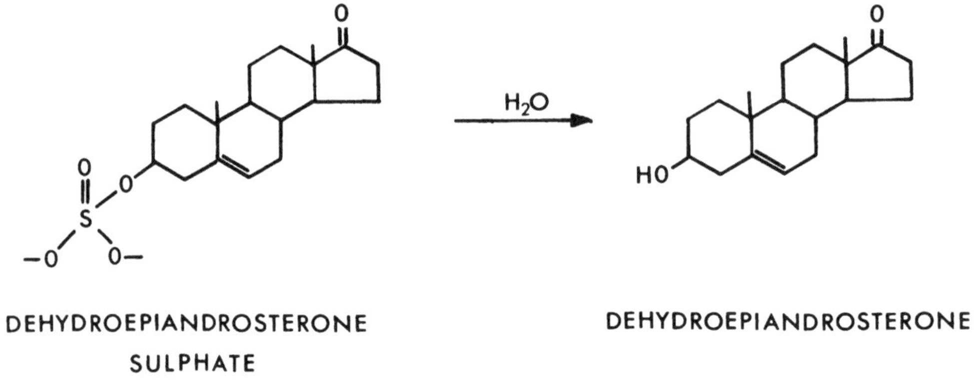

Fig. 13 Reaction catalyzed by steroid sulphatase

former tissue oestrone sulphatase has been studied and it is thought to be identical to arylsulphatase C but distinct from other steroid sulphatases (239). In the placenta, steroid sulphatase activity has been solubilized and partially purified (240). The sulphatase enzyme complex may be composed of subunits with different specificities. Earlier studies on the placental activity also indicated that multiple sulphatases were present (241, 242).

When the sulphatase reaction is carried out in the presence of $H_2{}^{18}O$, ^{18}O is found in the sulphate (243). This indicates that the O-S bond is broken during the reaction. The substrate specificity and ability of various steroids to inhibit the reaction has been studied with subcellular fractions from testis and human placenta (244-247). In the latter tissue it has been proposed that this inhibition is involved in the control of oestrogen synthesis (246, 247). Lipids have been found to affect steroid sulphatase activity in placental microsomes and disturbances in lipid metabolism are thought to be responsible for steroid sulphatase deficiency (248, 249).

This enzyme activity can be assayed by using radioactive

substrates and following the formation of the water-insoluble product, which is extracted into an organic solvent.

Conclusion

The understanding of the biosynthesis and metabolism of steroid hormones is of primary importance for the clinician since the clinical entities which result from an enzymatic defect in these processes are numerous. Active steroids are formed in specific organs and transformed into less active or totally inert substances and rendered water soluble to facilitate their excretion into the bile or urine. All of these transformations are catalyzed by specific enzyme systems.

From the preceding pages it is obvious that enzymatic defects in steroid metabolism can result from a number of factors. For example, recent studies on human placental sulphatase deficiency have indicated that this may be due to changes in the lipids present in the endoplasmic reticulum (248, 249). Also, it is possible that in hydroxylase systems, defects may result from abnormalities in the flavoproteins which are intermediate in the transfer of reducing equivalents from NADPH to the cytochrome P450's.

Acknowledgements

The expert assistance of N. Nérette and E. Gibb in the preparation of this article is gratefully acknowledged. W.G. is a scholar of the Fonds de la Recherche en Santé du Québec.

References

1. Klingenberg, M.: Pigments of rat liver microsomes. Arch. Biochem. Biophys. 75, 376-386 (1958).

2. Garfinkel, D.: Studies on pig liver microsomes. I. Enzymic and pigment composition of different microsomal fractions. Arch. Biochem. Biophys. 77, 493-509 (1958).

3. Omura, T., Sato, R.: The carbon monoxide-binding pigment of liver microsomes. I. Evidence for its hemoprotein nature. J. biol. Chem. 239, 2370-2378 (1964).

4. Hodgson, E.: Comparative aspects of the distribution of cytochrome P-450 dependent mono-oxygenase systems. Drug Metab. Rev. 10, 15-33 (1979).

5. Schenkman, J.B., Remmer, H., Eastabrook, R.W.: Spectral studies of drug interaction with hepatic microsomal cytochrome. Mol. Pharmacol. 3, 113-123 (1967).

6. White, R.E., Coon, M.J.: Oxygen activation by cytochrome P-450. Ann. Rev. Biochem. 49, 315-356 (1980).

7. Testa, B., Jenner, P.: Inhibitors of cytochrome P-450 and their mechanism of action. Drug Metab. Rev. 12, 1-117 (1981).

8. Thompson, E.A.Jr., Siiteri, P.K.: The involvement of human placental microsomal cytochrome P-450 in aromatization. J. biol. Chem. 249, 5373-5378 (1974).

9. Paine, A.J.: Hepatic cytochrome P-450. Essays Biochem. 17, 85-126 (1981).

10. Gillette, J.R.: Effects of induction of cytochrome P-450 enzymes on the concentration of foreign compounds and their metabolites and on the toxicological effects of these compounds. Drug. Metab. Rev. 10, 59-87 (1979).

11. Simpson, E.R., Boyd, G.S.: The cholesterol side-chain cleavage system of bovine adrenal cortex. Europ. J. Biochem. 2, 275-285 (1967).

12. Simpson, E.R.: Cholesterol side-chain cleavage, cytochrome P-450, and the control of steroidogenesis. Mol. Cell. Endocr. 13, 213-227 (1979).

13. Churchill, P.F., de Alvare, L.R., Kimura, T.: Topological studies of the steroid hydroxylase complexes in bovine adrenocortical mitochondria. J. biol. Chem. 253, 4924-4929 (1978).

14. Churchill, P.F., Kimura, T.: Topological studies of cytochromes P-450_{scc} and P-$450_{11\beta}$ in bovine adrenocortical inner mitochondrial membranes. J. biol. Chem. 254, 10443-10448 (1979).

15. Simpson, E.R., Boyd, G.S.: Partial resolution of the mixed-function oxidase involved in the cholesterol side-chain cleavage reaction in bovine adrenal mitochondria. Biochem. biophys. Res. Commun. 28, 945-950 (1967).

16. Bryson, M.J., Sweat, M.L.: Cleavage of cholesterol side-chain associated with cytochrome P-450, flavoprotein and non heme iron-protein derived from the bovine adrenal cortex. J. biol. Chem. 243, 2799-2804 (1968).

17. Takemori, S., Suhara, K., Hashimoto, K., Hashimoto, M., Sato, H., Gomi, T., Katagiri, M.: Purification of cytochrome P-450 from bovine adrenocortical mitochondria by an "aniline-sepharose" and the properties. Biochem. biophys. Res. Commun. 63, 588-593 (1975).

18. Wang, H.P., Kimura, T.: Purification and characterization of adrenal cortex mitochondrial cytochrome P-450 specific for cholesterol side-chain cleavage activity. J. biol. Chem. 251, 6068-6074 (1976).

19. Tilley, B.E., Watanuki, M., Hall, P.F.: Preparation and properties of side-chain cleavage cytochrome P-450 from bovine adrenal cortex by affinity chromatography with pregnenolone as ligand. Biochim. biophys. Acta 493, 260-271 (1977).

20. Kashiwagi, K., Dafeldecker, P., Salhanick, H.A.: Purification and characterization of mitochondrial cytochrome P-450 associated with cholesterol side-chain cleavage from bovine corpus luteum. J. biol. Chem. 255, 2606-2611 (1980).

21. Takikawa, O., Gomi, T., Suhara, K., Itagaki, E., Takemori, S., Katagiri, M.: Properties of an adrenal cytochrome P-450 ($P\text{-}450_{scc}$) for the side-chain cleavage of cholesterol. Arch. Biochem. Biophys. 190, 300-306 (1978).

22. Kido, T., Arakawa, M., Kimura, T.: Adrenal cortex mitochondrial cytochrome P-450 specific to cholesterol side-chain cleavage reaction. J. biol. Chem. 254, 8377-8385 (1979).

23. Hall, P.F., Watanuki, M., Degroot, J., Rouser, G.: Composition of lipids bound to pure cytochrome P-450 of cholesterol side-chain cleavage enzyme from bovine adrenocortical mitochondria. Lipids 14, 148-151 (1979).

24. Hall, P.F., Watanuki, M., Hamkalo, B.A.: Adrenocortical cytochrome P-450 side-chain cleavage. Preparations of membrane-bound side-chain cleavage system from purified components. J. biol. Chem. 254, 547-552 (1979).

25. Suhara, K., Gomi, T., Sato, H., Itagaki, E., Takemori, S., Katagiri, M.: Purification and immunochemical characterization of the two adrenal cortex mitochondrial cytochrome P-450-proteins. Arch. Biochem. Biophys. 190, 290-299 (1978).

26. Watanuki, M., Granger, G.A., Hall, P.F.: Cytochrome P-450 from bovine adrenocortical mitochondria immunochemical properties and purity. J. biol. Chem. 253, 2927-2931 (1978).

27. Kashiwagi, K., Carraway, R.E., Salhanick, H.A.: Amino acid composition of cytochrome $P\text{-}450_{scc}$ from bovine corpus luteum. Biochem. biophys. Res. Commun. 105, 110-116 (1982).

28. Nabi, N., Kominami, S., Takemori, S., Omura, T.: *In vitro* synthesis of mitochondrial cytochromes $P\text{-}450_{(scc)}$ and P-450 $_{(11\beta)}$ and microsomal cytochrome $P\text{-}450_{(C\text{-}21)}$ by both free and bound polysomes isolated from bovine adrenal cortex. Biochem. biophys. Res. Commun. 97, 687-693 (1980).

29. Dubois, R.N., Simpson, E.R., Kramer, R.E., Waterman, M.R.: Induction of synthesis of cholesterol side-chain cleavage cytochrome P-450 by adrenocorticotropin in cultured bovine adrenocortical cells. J. biol. Chem. 256, 7000-7005 (1981).

30. Wolfson, A.J., Lieberman, S.: Evidence suggesting that more than one sterol side-chain cleavage enzyme system exists in mitochondria from bovine adrenal cortex. J. biol. Chem. 254, 4096-4100 (1979).

31. Arthur, J.R., Blair, H.A.F., Boyd, G.S., Mason, J.I., Suckling, K.E.: Oxidation of cholesterol and cholesterol analogues by mitochondrial preparations of steroid-hormone-producing tissue. Biochem. J. 158, 47-51 (1976).

32. Morisaki, M., Duque, C., Ikekawa, N., Shikita, M.: Substrate specificity of adrenocortical cytochrome $P\text{-}450_{scc}$. I. Effect of structural modification of cholesterol side-chain on pregnenolone production. J. Steroid Biochem. 13, 545-550 (1980).

33. Morisaki, M., Duque, C., Takane, K., Ikekawa, N., Shikita, M.: Substrate specificity of adrenocortical cytochrome P-450. II. Effect of structural modification of cholesterol A/B ring on their side-chain cleavage reaction. J. Steroid Biochem. 16, 101-105 (1982).

34. Craig, I.F., Mason, J.I., Suckling, K.E., Boyd, G.S.: The influence of the side-chain on sterol side-chain cleavage in rat adrenal glands. Biochim. biophys. Acta 711, 123-127 (1982).

35. Shikita, M., Hall, P.F.: The stoichiometry of the conversion of cholesterol and hydroxycholesterols to pregnenolone catalyzed by adrenal cytochrome P-450. Proc. Nat. Acad. Sci. U.S.A. 71, 1441-1445 (1974).

36. Burstein, S., Dinh, J., Co, N., Gut, M., Schleyer, H., Cooper, D.Y., Rosenthal, O.: Kinetic studies on substrate-enzyme interaction in the adrenal cholesterol side-chain cleavage system. Biochemistry 11, 2883-2891 (1972).

37. Burstein, S., Middleditch, B.S., Gut, M.: Enzymatic formation of (20R, 22R)-20, 22-dihydroxycholesterol from cholesterol and a mixture of $^{16}O_2$ and $^{18}O_2$. Biochem. biophys. Res. Commun. 61, 692-697 (1974).

38. Burstein, S., Gut, M., Han, G.: Kinetics and mechanism of conversion of cholesterol to pregnenolone in bovine adrenal cortex. Proc. 57th Annu. Meeting Endocr. Soc. p. 56 (1975).

39. Constantopoulos, G., Carpenter, A., Satoh, P.S., Tchen, T.T.: Formation of isocaproaldehyde in the enzymatic cleavage of cholesterol side-chain by adrenal extract. Biochemistry 5, 1650-1652 (1966).

40. Takemoto, C., Hideki, H., Sato, H., Tamaoki, B.I.: Fate of molecular oxygen required by endocrine enzymes for the side-chain cleavage of cholesterol. Biochim. biophys. Acta 152, 749-757 (1968).

41. Hochberg, R.B., McDonald, P.D., Ladany, S., Lieberman, S.: Transient intermediates in steroidogenesis. J. Steroid Biochem. 6, 323-328 (1975).

42. Hochberg, R.B., Van der Hoeven, T.A., Welch, M., Lieberman, S.: A simple and precise assay of the enzymatic conversion of cholesterol into pregnenolone. Biochemistry 13, 603-609 (1974).

43. Rabe, T., Rabe, D., Bierwirth, A-M., Runnebaum, B.: New assays for the enzymatic conversion of cholesterol to pregnenolone. Steroids 37, 555-571 (1981).

44. Simpson, E.R., Miller, D.A.: Cholesterol side-chain cleavage, cytochrome P-450 and iron-sulfur protein in human placental mitochondria. Arch Biochem. Biophys. 190, 800-808 (1978).

45. Mitani, F.: Cytochrome P-450 in adrenocortical mitochondria. Mol. Cell. Biochem. 24, 21-43 (1979).

46. Sholl, S.A.: Estrogen synthetase and C^{17-20} lyase activity in fetal monkey gonad. J. Steroid Biochem. 16, 141-145 (1982).

47. Shibusawa, H., Sano, Y., Yoshida, N., Okinaga, S., Arai, K.: Studies of the human fetal adrenal gland - Properties of 17α-hydroxylase and C_{17}-C_{20} lyase in the biosynthesis of dehydroepiandrosterone from pregnenolone. J. Steroid Biochem. 9, 1125-1132 (1978).

48. Yoshida, K.I., Oshima, H., Troen, P.: Studies of the human testis. XIII. Properties of nicotinamide adenine dinucleotide-linked 17α-hydroxylation. J. clin. Endocrinol. Metab. 50, 895-899 (1980).

49. Hosaka, M., Oshima, H., Troen, P.: Studies of human testis. XIV. Properties of C17-C20 lyase. Acta endocrinologica 94, 389-396 (1980).

50. Inano, H., Tamaoki, B-I.: *In vitro* effect of 16α-hydroxyprogesterone on the enzyme activities related to androgen production in human testes. Acta endocrinologica 88, 768-777 (1978).

51. Nakano, H., Sato, H., Tamaoki, B-I.: On side-chain cleavage of 17α-hydroxyprogesterone by testicular enzyme. Biochem. biophys. Res. Commun. 22, 425-430 (1966).

52. Betz, G., Michels, D.: The effect of metyrapone on the 17, 20-lyase from rat testis microsomes. Biochem. biophys. Res. Commun. 50, 134-139 (1973).

53. Menard, R.H., Purvis, J.L.: Studies of cytochrome P-450 in testis microsomes. Arch Biochem. Biophys. 154, 8-18 (1973).

54. McMurtry, R.J., Hagerman, D.D.: Carbon monoxide inhibition of progesterone hydroxylation and side-chain cleavage catalyzed by rat testis microsomes. Steroids Lipids Res. 3, 8-13 (1972).

55. Nakajin, S., Shikita, M., Hall, P.F.: Microsomal cytochrome P-450 from neonatal pig testis: Purification and properties of a C_{21} steroid side-chain cleavage system. J. biol. Chem. 256, 3871-3876 (1981).

56. Betz, G., Roper, M., Tsai, P.: Steroid 17,20 lyase: Participation of NADPH cytochrome C reductase. Arch Biochem. Biophys. 163, 318-323 (1974).

57. Betz, G., Tsai, P., Hales, D.: Reconstitution of steroid 17,20 lyase activity after separation and purification of cytochrome P-450 and its reductase from rat testis microsomes. Endocrinology 107, 1055-1060 (1980).

58. Betz, G., Tsai, P., Weakley, R.: Heterogeneity of cytochrome P-450 in rat testis microsomes. J. biol. Chem. 251, 2839-2841 (1976).

59. Hochberg, R.B., Ladany, S., Lieberman, S.: Conversion of a C-20-deoxy-C21 steroid, 5-pregnen-3β-Ol, into testosterone by rat testicular microsomes. J. biol. Chem. 251, 3320-3325 (1976).

60. Nakajin, S., Shively, J.E., Yuan, P-M., Hall, P.F.: Microsomal cytochrome P-450 from neonatal pig testis: Two enzymatic activities (17α-hydroxylase and $C_{17,20}$ lyase) associated with one protein. Biochemistry 20, 4037-4042 (1981).

61. Nakajin, S., Hall, P.F., Onoda, M.: Testicular microsomal cytochrome P-450 for C_{21} steroid side-chain cleavage. J. biol. Chem. 256, 6134-6139 (1981).

62. Onoda, M., Hall, P.F.: Inhibition of testicular microsomal cytochrome P-450 (17α-hydroxylase/C-17,20 lyase) by estrogens. Endocrinology 109, 763-767 (1981).

63. Nakajin, S., Hall, P.F.: Side-chain cleavage of C_{21} steroids to C_{19} steroids by testicular microsomal cytochrome P-450. J. Steroid Biochem. 14, 1249-1253 (1981).

64. Barbieri, R.L., Canick, J.A., Ryan, K.J.: High-affinity steroid binding to rat testis 17α-hydroxylase and human placental aromatase. J. Steroid Biochem. 14, 387-393 (1981).

65. Betz, G., Tsai, P.: 17, 20 Lyase - Analysis of the cytochrome-P-450 binding site using multiple inhibitors. J. Steroid Biochem. 10, 393-395 (1979).

66. Betz, G., Michels, D.: Steroid 17,20-lyase: The effects of detergents on enzymic activity and microsomal composition. Steroids 21, 785-800 (1973).

67. Kremers, P.: New assay for pregnenolone and progesterone 17α-hydroxylase based on the specific substitution of a tritium situated on carbon 17. Europ. J. Biochem. 61, 481-486 (1976).

68. Cheatum, S.G., Warren, J.C.: Purification and properties of 3β-hydroxysteroid dehydrogenase and Δ^5-3-ketosteroid isomerase from bovine corpora lutea. Biochim. biophys. Acta 122, 1-13 (1966).

69. Cheatum, S.G., Douville, A.W., Warren, J.C.: Site specificity of bovine adrenal 3β-hydroxysteroid dehydrogenase and Δ^5-3-ketosteroid isomerase. Biochim. biophys. Acta 137, 172-178 (1967).

70. Koide, S.S., Torres, M.T.: Distribution of 3β-hydroxysteroid dehydrogenase and delta-5-3-oxosteroid isomerase in homogenate fractions of human term placenta. Biochim. biophys. Acta 105, 115-120 (1965).

71. Basch, R.S., Finegold, M.J.: 3β-Hydroxysteroid dehydrogenase activity in the mitochondria of rat adrenal homogenates. Biochem. J. 125, 983-989 (1971).

72. Gibb, W., Hagerman, D.D.: The specificity of the 3β-hydroxysteroid dehydrogenase activity of bovine ovaries toward dehydroepiandrosterone and pregnenolone. Steroids 28, 31-41 (1976).

73. Jackanicz, T.M., Armstrong, D.T.: Progesterone biosynthesis in rabbit ovarian interstitial tissue mitochondria. Endocrinology 83, 769-776 (1968).

74. Sulimovici, S., Boyd, G.S.: The Δ^5-3β-hydroxysteroid dehydrogenase of rat ovarian tissue. Europ. J. Biochem. 7, 549-558 (1969).

75. Ferre, F., Breuiller, M., Cedar, L., Duchesne, M.-J., Saintot, J., Descomps, B., Crastes de Paulet, A.: Human placental Δ^5-3β-hydroxysteroid dehydrogenase activity. Steroids 26, 551-570 (1975).

76. McCune, R.W., Roberts, S., Young, P.L.: Competitive inhibition of adrenal Δ^5-3β-hydroxysteroid dehydrogenase and Δ^5-3-ketosteroid isomerase activities by adenosine 3'5'-monophosphate. J. biol. Chem. 245, 3859-3867 (1970).

77. Cowan, R.A., Giles, C.A., Grant, J.K.: The intramitochondrial distribution of the 3β-hydroxysteroid dehydrogenase-oxosteroid isomerase. J. Steroid Biochem. 5, 607-608 (1974).

78. Chapman, J.C., Sauer, L.A.: Intracellular localization and properties of 3β-hydroxysteroid dehydrogenase/isomerase in the adrenal cortex. J. biol. Chem. 254, 6624-6630 (1979).

79. Headon, D.R., Hsiao, J., Ungar, F.: The intracellular localization of adrenal 3β-hydroxysteroid dehydrogenase/Δ^5 isomerase by density gradient perturbation. Biochem. biophys. Res. Commun. 82, 1006-1012 (1978).

80. Gibb, W., Lavoie, J.C., Roux, J.F.: 3β-Hydroxysteroid dehydrogenase activity in human fetal membranes. Steroids 32, 365-372 (1978).

81. Weidenfeld, J., Ben-Uzillio, R., Ne'eman, Y.: Cofactor requirements of microsomal 3β-hydroxysteroid oxidoreductase, 5-ene-isomerase from human placenta. J. Steroid Biochem. 6, 1243-1246 (1975).

82. Koritz, S.B.: The conversion of pregnenolone to progesterone by small particles from rat adrenal. Biochemistry 3, 1098-1102 (1964).

83. Gibb, W.: Substrate and nucleotide specificity of placental microsomal 3β-hydroxysteroid dehydrogenase. Steroids 37, 23-32 (1981).

84. Kowal, J., Forchielli, E., Dorfman, R.I.: The Δ^5-3β-hydroxysteroid dehydrogenases of corpus luteum and adrenal II interaction of C_{19} and C_{21} substrates and products. Steroids 4, 77-100 (1964).

85. Handler, R.P., Bransome, E.D.: Guinea pig adrenal 3β-hydroxysteroid dehydrogenase. J. clin. Endocrinol. Metab. 29, 1117-1119 (1969).

86. Yates, J., Deshpande, N.: Evidence for the existence of a single 3β-hydroxysteroid dehydrogenase/$\Delta^{5,4}$-3-oxosteroid isomerase complex in the human adrenal gland. J. Endocr. 64, 195-196 (1975).

87. Yoshida, N.: Steroid specificity of human placental 5-ene-3β-hydroxysteroid oxidoreductase. Steroids 33, 9-21 (1979).

88. Blomquist, C.H., Kotts, C.E., Hakanson, E.Y.: Modification of the kinetic properties of 5-ene-3β-hydroxysteroid dehydrogenase of human placental microsomes by hydrogen peroxide and 2-mercaptoethanol. J. Steroid Biochem. 9, 685-690 (1978).

89. Gibb, W.: Kinetic analysis of the placental microsomal 3β-hydroxysteroid dehydrogenase activity. Steroids 33, 459-466 (1979).

90. Ohba, H., Inano, H., Tamaoki, B-I.: Contribution of microsomal cytochrome b_5 electron transport system coupled with Δ^5-3β-hydroxysteroid dehydrogenase to androgen synthesis from progesterone. Biochem. biophys. Res. Commun. 103, 1273-1280 (1981).

91. Townsley, J.D.: Inhibition of placental 3β-hydroxysteroid dehydrogenase by naturally occurring steroids. Acta Endocrinol. (Kbh) 79, 740-748 (1975).

92. Wiener, M.: Control of placental 3β-hydroxy-Δ^5-steroid dehydrogenase: Comparison of enzyme characteristics in man, goat, rat and rhesus monkey. Biol. Reprod. 14, 306-313 (1976).

93. Wiener, M., Allen, S.H.G.: Inhibition of placental steroid synthesis by steroid metabolites. Steroids 12, 261-273 (1968).

94. Wiener, M., Allen, S.H.G.: Inhibition of placental steroid synthesis by steroid metabolites. Steroids 9, 567-581 (1967).

95. Gibb, W., Shapiro, S.S., Hagerman, D.D.: Inability to isolate the 3β-hydroxysteroid dehydrogenase from bovine ovaries

using a procedure which isolates the adrenal enzyme. J. Steroid Biochem. 7, 633-634 (1976).

96. Ford, H.C., Engel, L.L.: Purification and properties of the Δ^5-3β-hydroxysteroid dehydrogenase-isomerase system of sheep adrenal cortical microsomes. J. biol. Chem. 249, 1363-1368 (1974).

97. Wiebe, J.P., Larner, J.M.: A simple procedure for solubilizing 3β-hydroxysteroid dehydrogenase from microsomes of rat adrenals and testis interstitial cells. Steroids 30, 691-702 (1977).

98. Gallay, J., Vincent, M., De Paillerets, C., Alfsen, A.: Solubilization and separation of Δ^5-3β-hydroxysteroid dehydrogenase and 3-oxosteroid-Δ^4-Δ^5 isomerase from bovine adrenal cortex microsomes. Biochim. biophys. Acta 529, 79-87 (1978).

99. Penning, T.M., Covey, D.F., Talalay, P.: Irreversible Inactivation of Δ^5-3-ketosteroid isomerase of pseudomonas testosteroni by acetylenic suicide substrates. Mechanism of formation and properties of the steroid-enzyme adduct. J. biol. Chem. 256, 6842-6850 (1981).

100. Pollack, R.M., Kayser, R.H., Bevins, C.L.: An active-site-directed irreversible inhibitor of Δ^5-3-ketosteroid isomerase. Biochem. biophys. Res. Commun. 91, 783-790 (1979).

101. Ferrez, M., Renoir, M., Alfsen, A., Baulieu, E.E. Weintraub, H.: Antibodies to pseudomonas testosteroni 3-oxosteroid Δ^4 -Δ^5-isomerase. Biochem. biophys. Res. Commun. 95, 1238-1244 (1980).

102. Weintraub, H., Baulieu, E.-E., Alfsen, A.: Influence of the position of the double bond in steroid substrates on the efficiency of the proton-transfer reaction by pseudomonas testosteroni 3-oxosteroid Δ^4-Δ^5-isomerase. Biochem. J. 185, 723-732 (1980).

103. Penning, T.M., Covey, D.F., Talalay, P.: Inactivation of Δ^5-3-oxo steroid isomerase with active-site-directed acetylenic steroids. Biochem. J. 193, 217-227 (1981).

104. Penning, T.M., Talalay, P.: Linkage of an acetylenic secosteroid suicide substrate to the active site of Δ^5-3-ketosteroid isomerase isolation and characterization of a tetrapeptide. J. biol. Chem. 256, 6851-6858 (1981).

105. Smith, S.B., Richards, J.W., Benisek, W.F.: The purification and characterization of Δ^5-3-ketosteroid isomerase from pseudomonas putida, a cystine-containing isomerase. J. biol. Chem. 255, 2678-2684 (1980).

106. Smith, S.B., Benisek, W.F.: Active site-directed photoinactivation of Δ^5-3-ketosteroid isomerase from pseudomonas putida dependent on 1,4,6-androstrarien-3-one-17β-Ol. J. biol. Chem. 255, 2690-2693 (1980).

107. Bevins, C.L., Kayser, R.H., Pollack, R.M., Ekiko, D.B., Sadoff, S.: Irreversible active-site-directed inhibition of Δ^5-3-ketosteroid isomerase by steroidal 17β-oxiranes. Biochem. biophys. Res. Commun. 95, 1131-1137 (1980).

108. Rosner, J.M., Hall, P.F., Eik-Nes, K.B.: Conversion of progesterone to pregnenolone by rabbit testis. Steroids 5, 199-210 (1965).

109. Ward, M.G., Engel, L.L.: Reversibility of steroid Δ-isomerase. II. The reaction sequence in the conversion of androst-4-ene-3, 17-dione to 3β-hydroxyandrost-5-en-17-one by sheep adrenal microsomes. J. biol. Chem. 241, 3147-3153 (1966).

110. Shapiro, M.I.: Conversion of 17α, 21-dihydroxy-4-pregnene-3, 20-dione-4-^{14}C to 3β, 17α, 21-trihydroxy-5-pregnen-20-one-^{14}C by reversal of the 3β-hydroxysteroid dehydrogenase-isomerase reactions. Steroids 20, 1-12 (1972).

111. Ewald, W., Werbin, H., Chaikoff, I.L.: Evidence for two substrate specific Δ^5-3-ketosteroid isomerases in beef adrenal glands and their separation from 3β-hydroxysteroid dehydrogenase. Biochim. biophys. Acta 81, 199-201 (1964).

112. Ewald, W., Werbin, H., Chaikoff, I.L.: Evidence for the presence of 17-hydroxypregnenedione-isomerase in beef adrenal cortex. Biochim. biophys. Acta 111, 306-312 (1965).

113. Kruskemper, H.L., Forchielli, E., Ringold, H.J.: Δ^5-3-keto isomerases - Specificity and inhibition studies in bovine adrenal homogenate fractions and distribution in rat tissues. Steroids 3, 295-309 (1964).

114. Viger, A., Coustal, S., Marquet, A.: A reinvestigation of the mechanism of Δ^5-3-ketosteroid isomerase from bovine adrenal glands. Biochim. biophys. Acta 660, 170-173 (1981).

115. Akhtar, M., Calder, M., Smith, T., Wright, J.N.: Mechanistic and stereochemical studies on 3-oxosteroid Δ^4-Δ^5-isomerase from human placenta. Biochem. J. 185, 411-421 (1980).

116. Bertolino, A., Benson, A.M., Talalay, P.: Activation of Δ^5-3-ketosteroid isomerase of bovine adrenal microsomes by serum albumins. Biochem. biophys. Res. Commun. 88, 1158-1166 (1979).

117. Ryan, K.J.: Biological aromatization of steroids. J. biol. Chem. 234, 268-272 (1959).

118. Canick, J.A., Ryan, K.J.: Properties of the aromatase system associated with the mitochondrial fraction of human placenta. Steroids 32, 499-509 (1978).

119. Renwick, A.G.C., Oliver, J.F.: The aromatization of androstenedione by human placental mitochondria. Steroids 22, 123-132 (1973).

120. Bellino, F.L., Osawa, Y.: Localization of estrogen synthetase in the chorionic villus fraction after homogenization of human term placenta. J. clin. Endocrinol. Metab. 44, 699-707 (1977).

121. Thompson, E.A. Jr., Siiteri, P.K.: Partial resolution of the placental microsomal aromatase complex. J. Steroid Biochem. 7, 635-639 (1976).

122. Bellino, F.L., Osawa, Y.: Solubilization of estrogen synthetase from human term placental microsomes using detergents. J. Steroid Biochem. 9, 219-228 (1978).

123. Osawa, Y., Higashiyama, T.: Isolation of human placental cytochrome P-450 and its mechanism of action on androgen aromatization. In: "Microsomes, Drug Oxidations and Chemical Carcinogenesis", Eds. Coon, M.J., Conney, A.H., Estabrook, R.W., Gelbain, H.V., Gillette, J.R., O'Brien, J., Academic Press, New York, pp. 225-228 (1980).

124. Siiteri, P.K., Thompson, E.A. Jr.: Studies of human placental aromatase. J. Steroid Biochem. 6, 317-322 (1975).

125. Osawa, Y., Higashiyama, T., Nakamura, T.: Species specificity of estrogen biosynthesis in pregnancy. J. Steroid Biochem. 15, 449-453 (1981).

126. Meigs, R.A., Ryan, K.J.: Enzymatic aromatization of steroids. J. biol. Chem. 246, 83-87 (1971).

127. Canick, J., Ryan, K.J.: Cytochrome P-450 and the aromatization of 16α-hydroxytestosterone and androstenedione by human placental microsomes. Mol. Cell. Endocrinol. 6, 105-115 (1976).

128. Kautsky, M.P., Hagerman, D.D.: Kinetic properties of steroid 19-hydroxylase and estrogen synthetase from porcine ovary microsomes. J. Steroid Biochem. 13, 1283-1290 (1980).

129. Gibb, W., Lavoie, J.C.: Substrate specificity of the placental microsomal aromatase. Steroids 36, 507-519 (1980).

130. Thompson, E.A. Jr., Siiteri, P.K.: Utilization of oxygen and reduced nicotinamide adenine dinucleotide phosphate by human placental microsomes during aromatization of androstenedione. J. biol. Chem. 249, 5364-5372 (1974).

131. Kelly, W.G., Judd, D., Stolee, A.: Aromatization of Δ^4-androstene-3,17-dione, 19-hydroxy-Δ^4-androstene-3,17-dione and 19-oxo-Δ^4-androstene-3,17-dione at a common catalytic site in human placental microsomes. Biochemistry 16, 140-145 (1977).

132. Fishman, J., Goto, J.: Mechanism of estrogen biosynthesis-participation of multiple enzyme sites in placental aromatase hydroxylations. J. biol. Chem. 256, 4466-4471 (1981).

133. Longchampt, J.E., Gaul, C., Ehrenstein, M., Dorfman, R.I.: 19-Hydroxy-Δ^4-androstene-3,17-dione: An intermediate in estrogen biosynthesis. Endocrinology 66, 416-419 (1960).

134. Morato, T., Hayano, M., Dorfman, R.I., Axelrod, L.R.: The intermediate steps in the biosynthesis of estrogens from androgens. Biochem. biophys. Res. Commun. 6, 334-338 (1961).

135. Wilcox, R.B., Engel, L.L.: Kinetic studies on the role of 19-hydroxyandrost-4-ene-3,17-dione in estrogen biosynthesis. Steroids (suppl.) 1, 49-57 (1965).

136. Akhtar, M., Skinner, J.M.: The stereospecific removal of a C-19 hydrogen atom in oestrogen biosynthesis. Biochem. J. 114, 75-81 (1969).

137. Akhtar, M., Skinner, J.M.: The intermediary role of a 19-oxoandrogen in the biosynthesis of oestrogen. Biochem. J. 109, 318-321 (1968).

138. Meyer, A.S.: Conversion of 19-hydroxy-Δ^4-androstene-3,17-dione to estrone by endocrine tissues. Biochim. biophys. Acta 17, 441-442 (1955).

139. Braselton, W.E.Jr., Orr, J.C., Engel, L.L.: Identification by gas chromatography-mass spectrometry of modified C_{19} steroids by human placental microsomes. Steroids 24, 411-433 (1974).

140. Daux, W.L., Osawa, Y.: Steroid structure and function. VI. The molecular conformation of 19-hydroxy-4-androstene-3, 17-dione as an intermediate for estrogen synthetase. J. Steroid Biochem. 13, 383-386 (1980).

141. Akhtar, M., Calder, M.R., Corina, D.L., Wright, J.N.: Mechanistic studies on C-19 demethylation in oestrogen biosynthesis. Biochem. J. 201, 569-589 (1982).

142. Townsley, J.D., Brodie, H.J.: Studies on the mechanism of estrogen biosynthesis. III. The stereochemistry of aromatization of C_{19} and C_{18} steroids. Biochemistry 7, 33-40 (1968).

143. Fishman, J., Raju, M.S.: Mechanism of estrogen biosynthesis. Stereochemistry of C-1 hydrogen elimination in the aromatization of 2β-hydroxy-19-oxo-androstenedione. J. biol. Chem. 256, 4472-4477 (1981).

144. Goto, J., Fishman, J.: Participation of a non enzyme transformation in the biosynthesis of estrogens from androgens. Science 195, 80-81 (1977).

145. Hosada, H., Fishman, J.: Unusually facile aromatization of 2β-hydroxy-19-oxo-4-androstene-3,17-dione to estrone. J. Am. Chem. Soc. 96, 7325-7329 (1974).

146. Morand, P., Williamson, D.G., Layne, D.S., Lompa-Krzymien, P., Salvador, J.: Conversion of an androgen epoxide into 17β-estradiol by human placental microsomes. Biochemistry 14, 635-638 (1975).

147. Reed, K.C., Ohno, S.: Kinetic properties of human placental aromatase application of an assay measuring ^3H_2O release from 1β, 2β-^{3}H-androgens. J. biol. Chem. 251, 1625-1631 (1976).

148. Fishman, J., Guzik, H., Dixon, D.: Stereochemistry of estrogen biosynthesis. Biochemistry 8, 4304-4309 (1969).

149. Brodie, H.J., Raab, K., Possanza, G., Seto, N., Gut, M.: Further stereochemical studies of the catalytic reduction of $\Delta^{1,4}$-3-keto steroids with tritium. J. org. Chem. 34, 2697-2702 (1969).

150. Schade, K., Schubert, K.: Estrogen biosynthesis and its inhibition. Endokrinologie 74, 90-99 (1979).

151. Brodie, A.M.H.: Recent advances in studies on estrogen biosynthesis. J. Endocrinol. Invest. 2, 445-460 (1979).

152. Engel, L.L.: The biosynthesis of estrogens. In: "Handbook of Physiology", Section 7, Endocrinology, Eds. Greep, R.O., Astwood, E.B., American Physiological Society, Washington D.C., vol. 2, pp. 467-483 (1973).

153. Finkelstein, M., Shaefer, J.M.: Inborn errors of steroid biosynthesis. Physiol. Rev. 59, 353-406 (1979).

154. Suhara, K., Ikeda, Y., Takemori, S., Katagiri, M.: The purification and properties of NADPH-adrenodoxin reductase from bovine adrenocortical mitochondria. FEBS Lett. 28, 45-47 (1972).

155. Chu, J., Kimura, T.: Substrate-induced difference spectra, electron paramagnetic resonance and enzymatic properties of cholesterol depleted mitochondrial cytochrome P-450 of bovine adrenal cortex. J. biol. Chem. 248, 2089-2094 (1973).

156. Chu, J., Kimura, T.: Studies on adrenal steroid hydroxylases. J. biol. Chem. 248, 5183-5187 (1973).

157. Foster, R.P., Wilson, L.D.: Purification and characterization of adrenodoxin reductase from bovine adrenal cortex. Biochemistry 14, 1477-1484 (1975).

158. Sugiyama, T., Yamano, T.: Purification and crystallization of NADPH-adrenodoxin reductase from bovine adrenocortical mitochondria. FEBS Lett. 52, 145-148 (1975).

159. Hiwatashi, A., Ichikawa, Y., Maruya, N., Yamano, T., Aki, K.: Properties of crystalline reduced nicotinamide adenine dinucleotide phosphate-adrenodoxin reductase from bovine adrenocortical mitochondria. Biochemistry 15, 3082-3090 (1976).

160. Hiwatashi, A., Ichikawa, Y., Yamano, T., Maruya, N.: Properties of crystalline reduced nicotinamide adenine dinucleotide phosphate-adrenodoxin reductase from bovine adrenocortical mitochondria. Biochemistry 15, 3091-3097 (1976).

161. Lambeth, J.O., Kamin, H.: Adrenodoxin reductase. Properties of the complexes of reduced enzyme with $NADP^+$ and NADPH. J. biol. Chem. 251, 4299-4326 (1976).

162. Kimura, T.: Biochemical aspects of iron-sulphur linkage in non-heme iron protein with special reference to "Adrenodoxin" In: "Structure and Bonding", Eds. Jorgensen, C.K., Neiland, J.B., Nyholm, R.S., Reinen, D., Williams, R.J.P., Springer-Verlag, Berlin, Heidelberg, New York, vol. 5, pp. 1-28(1968).

163. Estabrook, R.W., Suzuki, K., Mason, J.I., Baron, J., Taylor, W.E., Simpson, E.R., Purvis, J., McCarthy, J.: Adrenodoxin: An iron-sulfur protein of adrenal cortex mitochondria. In: "Iron Sulphur Proteins", Ed. Lovenberg, W., Academic Press, New York and London, pp. 193-223 (1973).

164. Montelius, J., Carlsson, S., Lindstedt, J., Galaris, D., Höjeberg, B., Rydström, J.: Purification of adrenodoxin reductase from bovine adrenal cortex mitochondria by affinity chromatography. Adrenodoxin and membranous cytochrome $P\text{-}450_{11\beta}$. J. Steroid Biochem. 11, 1551-1559 (1979).

165. Seybert, D.W., Lambeth, J.D., Kamin, H.: The participation of a second molecule of adrenodoxin in cytochrome-P-450-catalyzed 11β hydroxylation. J. biol. Chem. 253, 8355-8358 (1978).

166. Lombardo, A., Defaye, G., Guidicelli, C., Monner, N., Chambaz, E.M.: Integration of purified adrenocortical cytochrome $P\text{-}450_{11\beta}$ into phospholipid vesicles. Biochem. biophys. Res. Commun. 104, 1638-1645 (1982).

167. Sato, H., Ashida, M., Suhara, K., Itagaki, E., Takemori, S., Katagira, M.: Properties of an adrenal cytochrome P-450 for the hydroxylations of corticosteroids. Arch. Biochem. Biophys. 190, 307-314 (1978).

168. Watanuki, M., Tilley, B.E., Hall, P.F.: Cytochrome 11-β and 18-hydroxylase activities of bovine adrenocortical mitochondria. One enzyme or two ? Biochemistry 17, 127-130 (1978).

169. Björkhem, I., Karlmar, K.E.: 18-Hydroxylation of deoxycorticosterone by reconstituted systems from rat and bovine adrenals. Europ. J. Biochem. 51, 145-154 (1975).

170. Ryan, K.J., Engel, L.L.: Hydroxylation of steroids at carbon 21. J. biol. Chem. 225, 103-114 (1957).

171. Cooper, D.Y., Levin, S., Narasimhula, S., Rosenthal, O., Estabrook, R.W.: Studies on the mechanism of C-21 hydroxylation of steroids by the adrenal cortex. In: "Functions of the Adrenal Cortex", Ed. McKerns, K.W., Appleton-Century-Crofts, New York, vol. 11, pp. 897-942 (1968).

172. Cooper, D.Y., Estabrook, R.W., Rosenthal, O.: The stoichiometry of C-21 hydroxylation of steroids by adrenocortical microsomes. J. biol. Chem. 238, 1320-1323 (1963).

173. Masters, B.S.S., Baron, J., Taylor, W.E., Isaacson, E.L., Lo Spalluto, J.: Immunochemical studies on the electron transport chains involving cytochrome P-450. J. biol. Chem. 246, 4143-4150 (1971).

174. Masters, B.S.S., Taylor, W.E., Isaacson, E.L., Baron, J., Harkins, J.B., Nelson, E.B., Bryan, G.T.: Studies on the function of adrenodoxin and TPNH-cytochrome C reductase in the mitochondria and microsomes of adrenal cortex utilizing immunochemical techniques. Ann. N.Y. Acad. Sci. 212, 76-88 (1972).

175. Hiwatashi, A., Ichikawa, Y.: Physicochemical properties of reduced nicotinamide adenine dinucleotide phosphate-cytochrome-P-450 reductase from bovine adrenocortical microsomes. Biochim. biophys. Acta 580, 44-63 (1979).

176. Kominami, S., Ochi, H., Kobayashi, Y., Takemori, S.: Studies on the steroid hydroxylation system in adrenal cortex microsomes. J. biol. Chem. 255, 3386-3394 (1980).

177. Hiwatashi, A., Ichikawa, Y.: Purification and reconstitution of the steroid 21-hydroxylase system (cytochrome P-450-linked mixed function oxidase system) of bovine adrenocortical microsomes. Biochim. biophys. Acta 664, 33-48 (1981).

178. Kominami, S., Mori, S., Takemori, S.: Purification and optical studies of cytochrome P-450 from bovine adrenocortical microsomes. FEBS Letts. 89, 215-218 (1978).

179. Chasalow, F.I., Lieberman, S.: The activation of microsomal steroid 21-hydroxylase by cytosol from the cortex of bovine adrenals. J. biol. Chem. 254, 3777-3781 (1979).

180. Ponticorvo, L., Greenfield, N., Wolfson, A., Chasalow, F., Lieberman, S.: The nature of the cytosolic activators of the adrenal steroid 21-hydroxylation system. Arch. Biochem. Biophys. 200, 223-231 (1980).

181. Ulick, S.: Diagnosis and nomenclature of the disorders of the terminal portion of the aldosterone biosynthetic pathway. J. clin. Endocrinol. Metab. 43, 92-96 (1976).

182. Nicholis, G.L., Ulick, S.: Role of 18-hydroxylation in the biosynthesis of aldosterone. Endocrinology 76, 514-521 (1965).

183. Pasqualini, J.R.: Conversion of tritiated 18-hydroxycorticosterone to aldosterone by slices of human cortico-adrenal gland and adrenal tumour. Nature 201, 501 (1964).

184. Raman, P.B., Sharma, D.C., Dorfman, R.I.: Studies on aldosterone biosynthesis *in vitro*. Biochemistry 5, 1795-1804 (1966).

185. Psychoyos, S., Tallan, H.H., Greengard, P.: Aldosterone synthesis by adrenal mitochondria. J. biol. Chem. 241, 2949-2956 (1966).

186. Ulick, S.: Normal and alternate pathways in aldosterone biosynthesis. In: Proc. 4th Int. Congr. Endocrinol., Washington, D.C. 1972, Excerpta Med. Found. Int. Congr. Ser. No. 273, Amsterdam, pp. 761-767 (1974).

187. Levine, L.S., Rauh, W., Gottesdiener, K., Chow, D., Gunczler, P., Rapaport, R., Pang, S., Schneider, B., New, M.I.: New studies of the 11β-hydroxylase and 18-hydroxylase enzymes in the hypertensive forms of congenital adrenal hyperplasia. J. clin. Endocrinol. Metab. 50, 258-263 (1980).

188. Lehoux, J.G.: Factors controlling the biosynthesis of aldosterone. In: "Steroid Biochemistry", vol. 2, Ed. Hobkirk, R., C.R.C. Press, Boca Raton, Florida, pp. 51-80 (1979).

189. Tamaoki, B.: Steroidogenesis and cell structure. J. Steroid Biochem. 4, 89-118 (1973).

190. Ryan, K.J., Engel, L.L.: The interconversion of estrone and estradiol by human tissue slices. Endocrinology 52, 287-291 (1953).

191. Langer, L.J., Engel, L.L.: Human placental estradiol-dehydrogenase. 1 - Concentration, characterization and assay. J. biol. Chem. 233, 583-588 (1958).

192. Karavolos, H.J., Orr, J.C., Engel, L.L.: Human placental estradiol dehydrogenase. IV. Differentiation of 17β-estradiol-activated transhydrogenase from the transhydrogenase function of 17β-estradiol dehydrogenase. J. biol. Chem. 244, 4413-4421 (1969).

193. Engel, L.L., Groman, E.V.: Human placental oestradiol dehydrogenase. Recent Prog. Horm. Res. 30, 139-169 (1969).

194. Karavolos, H.J., Baedecker, M.L., Engel, L.L.: Human placental estradiol dehydrogenase. V. Purification and partial characterization of the diphosphopyridine nucleotide-linked enzyme. J. biol. Chem. 245, 4948-4952 (1970).

195. Nicolas, J.C., Pons, M., Descomps, B., Crastes de Paulet, A.: Affinity chromatography: Large scale purification of the soluble estradiol-dehydrogenase of human placenta. FEBS Lett. 23, 175-179 (1972).

196. Chin, C.C., Warren, J.C.: Purification of estradiol dehydrogenase from human placenta by affinity chromatography. Steroids 22, 373-378 (1973).

197. Chin, C.C., Dence, J.B., Warren, J.C.: Crystallization of human placental estradiol 17β-dehydrogenase. J. biol. Chem. 251, 3700-3705 (1976).

198. Jarabak, J., Street, M.A.: Studies on the soluble 17β-hydroxysteroid dehydrogenase from human placenta. Evidence for a subunit structure. Biochemistry 10, 3831-3834 (1971).

199. Burns, D.J.W., Engel, L.L., Bethune, J.L.: Amino acid composition and subunit structure: Human placental 17β-estradiol dehydrogenase. Biochemistry 11, 2699-2703 (1972).

200. Pons, M., Nicolas, J.C., Boussioux, A.M., Descomps, B., Crastes de Paulet, A.: Some new developments in the knowledge of human placental estradiol-17β dehydrogenase. J. Steroid Biochem. 8, 345-358 (1977).

201. Chin, C.C., Asmar, P., Warren, J.C.: Synthesis of 2-bromoacetamidoestrone methyl ether and study of the steroid binding site of human placental estradiol 17β-dehydrogenase. J. biol. Chem. 255, 3660-3664 (1980).

202. Inano, H., Engel, L.L.: Photoaffinity labelling of human placental estradiol dehydrogenase with 3-(arylazide-β-alanine) estrone. J. biol. Chem. 255, 7694-7699 (1980).

203. Pons, M., Nicolas, J.C., Boussioux, A.M., Descomps, B., Crastes de Paulet, A.: Affinity labelling of estradiol-17β dehydrogenase from human placenta with substrate analogs. Europ. J. Biochem. 68, 385-394 (1976).

204. Biellmann, J.F., Branlant, G., Nicolas, J.C., Pons, M., Descomps, B., Crastes de Paulet, A.: Alkylation of estradiol 17β-dehydrogenase from human placenta with 3-chloroacetylpyridine-adenine dinucleotide. Europ. J. Biochem. 63, 477-481 (1976).

205. Chin, C.C., Warren, J.C.: Synthesis of 16α-bromoacetoxyestradiol 3-methyl ether and study of the steroid binding site of human placental estradiol 17β-dehydrogenase. J. biol. Chem. 250, 7682-7686 (1975).

206. Strickler, R.C., Tobias, B., Covey, D.F.: Human placental 17β-estradiol dehydrogenase and 20α-hydroxysteroid dehydrogenase. Two activities at a single enzyme active site. J. biol. Chem. 256, 316-321 (1981).

207. Gibb, W., Jeffery, J.: Relationship between the 3α- and 20β-hydroxysteroid NAD-oxidoreductase activity of a crystalline-enzyme preparation. Europ. J. Biochem. 23, 336-342 (1971).

208. Sweet, F., Samant, B.R.: Bifunctional enzyme activity at the same active site: Study of 3α and 20β activity by affinity alkylation of 3α, 20β-hydroxysteroid dehydrogenase with 17-(bromoacetoxy) steroids. Biochemistry 19, 978-986 (1980).

209. Strickler, R.C., Covey, D.F., Tobias, B.: Study of 3α, 20β-hydroxysteroid dehydrogenase with an enzyme-generated affinity alkylator: Dual enzyme activity at a single active site. Biochemistry 19, 4950-4954 (1980).

210. Edwards, C.A.F., Orr, J.C.: Comparison of the 3α- and 20β-hydroxysteroid dehydrogenase activities of the cortisone reductase of Streptomyces Hydrogenans. Biochemistry 17, 4370-4376 (1978).

211. Jarabak, J., Talalay, P.: Stereospecificity of hydrogen transfer by pyridine nucleotide linked hydroxysteroid dehydrogenases. J. biol. Chem. 235, 2147-2151 (1960).

212. Warren, J.C., Gonsalez, M.C., Soria, J.: Characteristics of the steroid binding site: A method to evaluate stereospecificity of hydrogen transfer by steroid dehydrogenases. Endocrinology 80, 884-888 (1967).

213. Engel, L.L.: Eli Lilly Lecture: Estrogen metabolism and action. Endocrinology 87, 827-835 (1970).

214. Williamson, D.G.: Biochemistry of steroid 17-hydroxysteroid dehydrogenases. In: "Steroid Biochemistry", vol. 1, Ed. Hobkirk, R., C.R.C. Press Inc., Boca Raton, Florida, pp. 83-110 (1979).

215. Payne, A.H., Singer, S.S.: The role of steroid sulfatase and sulfotransferase enzymes in the metabolism of C_{21} and C_{19} steroids. In: "Steroid Biochemistry", vol. 1, Ed. Hobkirk, R., C.R.C. Press Inc., Boca Raton, Florida, pp. 111-145 (1979).

216. Adams, J.B., Ellyard, R.K., Low, J.: Enzymic synthesis of steroid sulphates. IX. Physical and chemical properties of purified oestrogen sulphotransferase from bovine adrenal glands. Biochim. biophys. Acta 370, 160-188 (1974).

217. Adams, J.B., Poulos, A.: Enzymic synthesis of steroid sulphates. III. Isolation and properties of estrogen sulphotransferase of bovine adrenal glands. Biochim. biophys. Acta 146, 493-508 (1967).

218. Adams, J.B., Low, J.: Enzymic synthesis of steroid sulphates. X. Isolation of oestrogen sulphotransferase from bovine placenta. Biochim. biophys. Acta 370, 189-196 (1974).

219. Horwitz, J.P., Misra, R.S., Rozhin, J., Helmer, S., Bhuta, A., Brooks, S.C.: Studies on bovine adrenal sulfotransferase. V. Synthesis and assay of analogs of 3'-phosphoadenosine 5'-phosphosulfate as cosubstrates for estrogen sulfuration. Biochim. biophys. Acta 613, 85-94 (1980).

220. Horwitz, J.P., Misra, R.S., Rozhin, J., Neenan, J.P., Huo, A., Godefroi, V.C., Philips, K.D., Chung, H.I., Butke, G., Brooks, S.C.: IV. Synthesis and assay of analogues of adenosine 3', 5'-diphosphate as inhibitors of bovine adrenal estrogen sulfotransferase. Biochim. biophys. Acta 525, 364-372 (1978).

221. Ryan, R.A., Carroll, J.: Studies on a 3β-hydroxysteroid sulphotransferase from rat liver. Biochim. biophys. Acta 429, 391-401 (1976).

222. Lyon, E.S., Jakoby, W.B.: The identity of alcohol sulfotransferases with hydroxysteroid sulfotransferases. Arch. Biochem. Biophys. 202, 474-481 (1980).

223. Adams, J.B., McDonald, D.: Enzymic synthesis of steroid sulphates. XII. Isolation of dehydroepiandrosterone sulphotransferase from human adrenals by affinity chromatography. Biochim. biophys. Acta 567, 144-153 (1979).

224. Adams, J.B., McDonald, D.: Enzymic synthesis of steroid sulphates. XIII. Isolation and properties of dehydroepiandrosterone sulphotransferase from human foetal adrenals. Biochim. biophys. Acta 615, 275-278 (1980).

225. Payne, A.H.: Testicular steroid sulfotransferase. Endocrinology 106, 1365-1370 (1980).

226. Bouthillier, M., Chapdelaine, A., Bleau, G., Roberts, K.D.: Steroid sulfotransferase in hamster epididymis. Steroids 38, 523-537 (1981).

227. Singer, S.S., Giera, D., Johnson, J., Sylvester, S.: Enzymatic sulfation of steroids: 1. The enzymatic basis for the sex difference in cortisol sulfation by rat liver preparations. Endocrinology 98, 963-974 (1976).

228. Singer, S.S., Moshtaghie, A., Lee, A., Kutzer, T.: Enzymatic sulfation of steroids. X. Pharmacological progesterone effects on rat liver glucocorticoid sulfotransferases. Biochem. Pharmacol. 29, 3181-3188 (1980).

229. Singer, S.S.: Enzymatic sulfation of steroids. III. The sulfation of corticosterone by the glucocorticoid sulfotransferase of rat liver cytosol. Biochim. biophys. Acta 539, 19-30 (1978).

230. Singer, S.S., Sylvester, S.: Enzymatic sulfation of steroids. II. The control of the hepatic cortisol sulfotransferase activity. Endocrinology 99, 1346-1352 (1976).

231. Singer, S.S.: Enzymatic sulfation of steroids. IV. Control of hepatic glucocorticoid sulfotransferase activity and the individual glucocorticoid sulfotransferases from male and female rats. Endocrinology 103, 66-73 (1978).

232. Singer, S.S., Kutzer, T., Lee, A.: Enzymatic sulfation of steroids. VIII. Control of hepatic cortisol sulfation and glucocorticoid sulfotransferases of rats by the pituitary gland. Endocrinology 104, 571-575 (1979).

233. Lewis, W.G., Witt, K.R., Singer, S.S.: Enzymatic sulfation of steroids. XIII. Hepatic deoxycorticosterone sulfotransferases in rats. Proc. Soc. exp. Biol. Med. 166, 70-75 (1981).

234. Singer, S.S., Burns, L.: Enzymatic sulfation of steroids. XI. The extensive purification of some properties of hepatic sulfotransferase from female rats. Can. J. Biochem. 58, 660-666 (1980).

235. Singer, S.S.: Partial purification and some properties of rat liver sulfotransferase. Arch. Biochem. Biophys. 196, 340-349 (1979).

236. Singer, S.S., Gebhart, J., Hess, E.: Enzymatic sulfation of steroids. V. Partial purification and some properties of sulfotransferase. Can. J. Biochem. 56, 1028-1035 (1978).

237. Sekura, R.D., Marcus, C.J., Lyon, E.S., Jakoby, W.B.: Assay of sulfotransferases. Anal. Biochem. 95, 82-86 (1979).

238. Jakoby, W.B., Sekura, R.D., Lyon, E.S., Marcus, C.J., Wang, J.L.: Sulfotransferases. In: "Enzymatic Basis of Detoxication", vol. II, Ed. Jakoby, W.B., Academic Press, New York, pp. 199-228 (1980).

239. Iwamori, M., Moser, H.W., Kishimoto, Y.: Solubilization and partial purification of steroid sulfatase from rat liver: Characterization of estrone sulfatase. Arch. Biochem. Biophys. 174, 199-208 (1976).

240. Gauthier, R., Vigneault, N., Bleau, G., Chapdelaine, A., Roberts, K.D.: Solubilization and partial purification of steroid sulfatase of human placenta. Steroids 31, 783-798 (1978).

241. French, A.P., Warren, J.C.: Properties of steroid sulphatase and arylsulphatase activities of human placenta. Biochem. J. 105, 233-241 (1967).

242. French, A.P., Warren, J.C.: Sulfatase activity in the human placenta. Steroids 8, 79-85 (1966).

243. Logan, G.G., Warren, J.C.: Hydrolysis of dehydroepiandrosterone sulphate by human placental steroid 3β-sulphatase in [^{18}O] water. Biochem. J. 114, 707-710 (1969).

244. Payne, A.H., Mason, M., Jaffe, R.B.: Testicular steroid sulfatase. I. Substrate specificity and inhibition. Steroids 14, 685-704 (1969).

245. Payne, A.H.: Gonadal steroid sulfates and sulfatase. V. Human testicular steroid sulfatase: Partial characterization and possible regulation by free steroids. Biochim. biophys. Acta 258, 473-479 (1972).

246. Townsley, J.D., Scheel, D.A., Rubin, E.J.: Inhibition of steroid 3-sulfate by endogenous steroids. J. clin. Endocrinol. Metab. 31, 670-678 (1970).

247. Townsley, J.D.: Further studies on the regulation of human placental steroid 3-sulfatase activity. Endocrinology 93, 172-181 (1973).

248. McKee, J.W.A., Abeysekera, R., France, J.T.: Studies of the biochemical basis of steroid sulphatase deficiency-11. A finding of decreased phospholipid content in sulphatase deficient placental microsomes. J. Steroid Biochem. 14, 195-198 (1981).

249. McNaught, R.W., France, J.T.: Studies of the biochemical basis of steroid sulphatase deficiency. J. Steroid Biochem. 13, 363-373 (1980).

MALE PSEUDOHERMAPHRODITISM DUE TO ABNORMAL STEROID SYNTHESIS, METABOLISM AND ACTION

P. Saenger
Department of Pediatrics, Montefiore Hospital and Medical Center
111 East 210th Street, Bronx, New York 10467, U.S.A.

Introduction

The remarkable advances in the field of sex differentiation and sex determination represent major landmarks in physiology. No aspect of prenatal development is better understood. This review is designed to help physicians evaluate sexually ambiguous patients by the selection of appropriate investigations and interpretation of the results.

Sex determination and differentiation is a sequential process that involves at least four characteristics which become delineated at various stages:

Genetic sex is determined at fertilization. Gonadal sex is determined by the genetic sex. Sexual differentiation of the gonads leads to the development of the internal genital tract and the external genitalia and hence, the phenotypic sex, during the first half of fetal life. At puberty, the development of sex specific secondary sex characteristics provides additional visible phenotypic manifestations.

Psychological sexual identity is acquired postnatally by sociological imprinting on the developing personality. This postnatal process together with prenatal hormonal influences (1), mould the gender identity, the psychological sex.

The term male pseudohermaphroditism (MPH) describes an individual with ambiguity of the internal or external genitalia whose karyotype is 46 XY and who has testes. MPH can be caused

Steroid Converting Enzymes and Diseases

by defects in testosterone synthesis, testosterone metabolism (5α-reductase deficiency) and androgen action. The purpose of this chapter is to describe enzymatic deficiencies of testosterone synthesis and metabolism resulting in male pseudohermaphroditism.

Normal Sexual Differentiation

Gonadal differentiation and the role of H-Y antigen

A gene or set of genes on the Y-chromosome causes the indifferent embryonic gonad to develop as a testis. A fetal ovary develops in the absence of an X-chromosome. In mammals, female sex differentiation is the innate tendency of the gonadal and genital primordia, on which maleness must be actively imposed (2). Once the critical period is over, by approximately 9 weeks of gestation, a genital primordium is irreversibly committed to femaleness and can no longer be induced to differentiate into a male gonad (Fig. 1).

In the 5mm fetus, the primitive gonad appears as a thickening of the coelomic epithelium covering the medioventral surface of the mesonephros, or primitive kidney. Shortly thereafter, the primitive gonad is colonized by primordial germ cells, travelling from the dorsal yolk sac (3, 4).

Development of testis and ovary. (Fig. 2). In the male gonadal ridge, under the influence of the Y-chromosome, the contact between primordial cells and somatic cells is made earlier than in the female gonadal ridge and leads to formation of seminiferous tubules rather than follicles (5). This step at about 45-50 days of gestation, is mediated by the gene product of the Y-chromosome, H-Y antigen (see below) and coincides with the acquisition of anti-Müllerian activity by the male gonad (6). Both H-Y antigen and anti-Müllerian hormone are secreted by the versatile Sertoli cell (5). Leydig cells appear at about

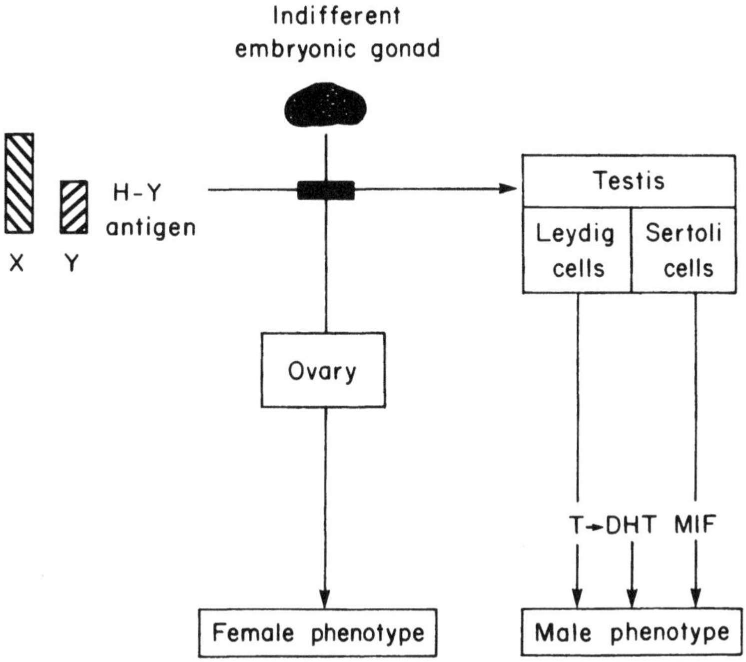

Fig. 1 Overview of sex differentiation. In normal XX embryos the indifferent embryonic gonad becomes an ovary and further development is female. In XY embryos H-Y antigen blocks female development by inducing the gonad to differentiate as a testis. Sertoli cells secrete Müllerian inhibiting factor (MIF). Leydig cells secrete testosterone which induces differentiation of the Wolffian derivatives. Some of the testosterone (T) is converted to dihydrotestosterone (DHT) which induces differentiation of external male genitalia. (From 96).

day 60 (3) and by 14-18 weeks make up more than half the volume of the testis (7). This coincides with the peak pituitary hormone chorionic gonadotropin (HCG) release. The fetal pituitary thus controls testicular development in the 2nd and 3rd trimes-

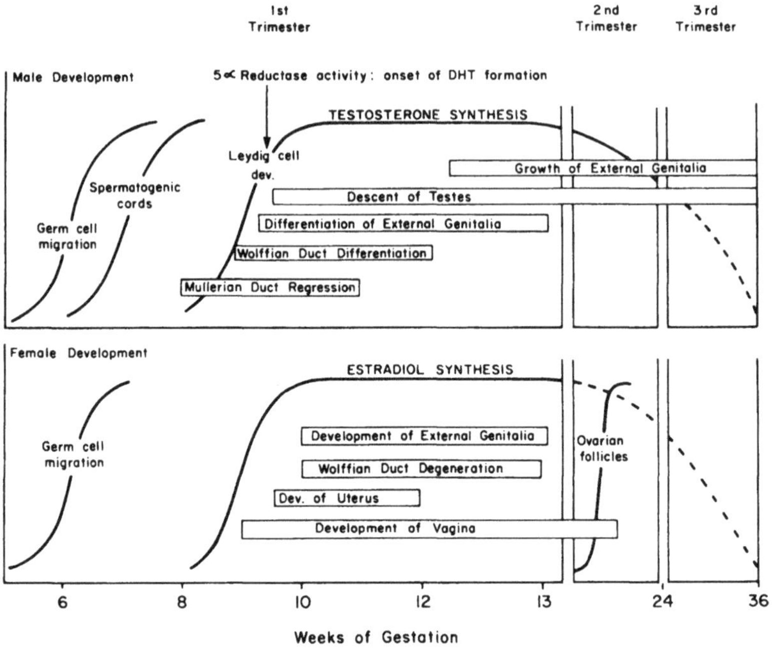

Fig. 2 Relation between differentiation of the gonads and the anatomical differentiation of the human male and female embryos. (Adapted from 8).

ter. Soon after the 18th week the number of Leydig cells decreases and by 27 weeks the seminiferous tubules are separated by only a narrow interstitium containing few Leydig cells. Several months after birth Leydig cells are not visible histologically; they reappear at puberty (3).

The onset of ovarian and testicular function in the human embryo occurs between 6 and 8 weeks of gestation (8, 14). Thus testes and ovaries acquire the capacity to secrete their characteristic hormones at the same stage of embryonic development. In the male, testosterone synthesis coincides with the histological appearance of Leydig cells and precedes the onset of virilization of the genital ducts.

The morphologic development of the fetal ovary follows a much slower time course. By 20 weeks of fetal age, more than three months after the start of testicular development, ovarian germ cell population reaches a peak of approximately 7 million germ cells. This population declines in the second half of pregnancy to 2 million at term as a result of cessation of mitotic activity and atresia (9).

Gonadotropins do not appear to be required for testosterone biosynthesis until late in embryogenesis (10). Late testosterone mediated events in male development, such as growth of the male external genitalia are modulated indirectly by hormones from the pituitary and the placenta. This explains the known clinical association of congenital hypopituitarism and microphallus (11).

The ontogeny of hormonal biosynthesis has been studied in mammalian gonads. In the rabbit embryo testosterone synthesis begins during a 12 hour period between days 17 and 17.5 of gestation approximately one day after the histological differentiation of the testis. This process is made possible by the appearance of 3β-hydroxysteroid dehydrogenase converting dehydroepiandrosterone to androstenedione. The ovary begins to form estradiol at the same time that the testis begins to synthesize testosterone; this occurs before histological differentiation of the ovary.

The onset of hormonal biosynthesis in the rabbit gonad is independent of the pituitary and of other hormonal control (12). Thus endocrine differentiation of the rabbit gonad can be studied in organ culture.

These findings suggest that while histologic differentiation of the indifferent embryonic gonad into a testes is controlled by H-Y antigen, the differentiation of the gonads into endocrine organs is controlled by factors intrinsic to the gonads. The mechanisms which convert the gonad from an autonomous endocrine tissue to a gonadotropin dependent tissue are not understood.

H-Y Antigen. The inducer substance for testicular organogenesis proposed by Jost (2) appears to be H-Y antigen. This antigen is the first plasma membrane or cell surface protein to which a specific organogenesis function has been assigned (13). The term H-Y (histocompatibility Y) antigen referred originally to a weak antigen that stimulated the rejection of male grafts by female mice of the same inbred strain. H-Y antigen can be measured semi-quantitatively in male serum and tissues because it is present on the surface of all male cells (14, 15). The recent success in establishing monoclonal anti H-Y antisera promises to facilitate analysis and clarification of the H-Y system (16). This clarification is needed because uncertainties about specificity and reproducibility of serologic tests for H-Y antigen exist. The often provocative and rapidly expanding literature on H-Y antigen assays, particularly when applied to clinical cases of abnormal sex differentiation, deserves critical evaluation. Although H-Y antigen appears to be the signal to testicular differentiation, it is also a histocompatibility factor. In other words, this single gene locus contributes to two of the most powerful differentiating attributes of individuals: male versus female, and self versus others (17).

Using the sperm cytotoxicity test, the invariant association of H-Y antigen with the heterogametic sex has been shown in a wide range of vertebrates, including mammals (18). H-Y antigen appears already in the eight cell stage of embryonic development (19), well before differentiation of gonadal structures. These observations lead Wachtel and Ohno to postulate that H-Y antigen is the factor responsible for inducing testicular organogenesis and that it is a product of the testis organizing gene (20). In support of this hypothesis is that XYY males express increased amounts of H-Y antigen (21). Chromosomal males with testicular feminization and a female phenotype are H-Y antigen positive (22). As the ultimate test of the organogenesis-directing capability of this antigen, it was shown that H-Y antigen isolated from Daudi Burkitt lymphoma induced precocious testicular transformation of female bovine undifferentiated gonads within 5 days of placement in organ culture. Conversely, exposure of a suspension of rat testes cells to excess anti H-Y serum lead to reaggregation of testicular cells to form ovarian follicles. These two types of experiments provide the fundamental and most cogent evidence in mammals for the testes organizing function of H-Y antigen (23, 24).

In summary, there is good evidence that H-Y antigen disseminated in a hormone-like fashion during embryogenesis (25) and bound by gonad specific receptors, appears to induce the indifferent XX or XY gonadal primordium to differentiate as a testis. In the absence of H-Y antigen or its specific receptor on the cell surface, the embryonic gonad follows its inherent tendency to differentiate into an ovary. Assay procedures available at present preclude, however, routine use of the test for H-Y antigen in newborns with ambiguous genitalia.

Differentiation of genital ducts (see Fig. 2)

The Müllerian ducts serve as the anlagen of the uterus, fallopian tubes and the upper part of the vagina, whereas the mesonephric or Wolffian ducts have the potential of differentiating further into epididymis, vas deferens, seminal vesicles and the ejaculatory ducts of the male. As shown by the classical experiments of Jost (2), using the rabbit fetus, degeneration of Müllerian ducts and differentiation of Wolffian ducts can be imposed on females only by the hormonal secretory products of the fetal testis. The pattern of testicular testosterone formation in the human embryo coincides closely with that for differentiation of the male urogenital tract, and the maximum placental HCG release (26).

This process is mediated by two discrete substances. Testosterone produced by fetal Leydig cells, virilizes and stabilizes Wolffian ducts. 5α-Dihydrotestosterone (DHT), acts on the urogenital sinus and the genital tubercle to complete normal masculine development of the external genitalia (27). Androgens exert their virilizing effect on the Wolffian duct system, the urogenital sinus and the external genitalia by binding to a cytosolic receptor specified by X chromosomal genes. This receptor exhibits a higher affinity toward DHT, than toward testosterone (28). The mesenchyme of the Wolffian duct system too exhibits androgen receptor activity and is an additional mediator of the morphogenetic effects of androgens upon the epithelium (29).

Anti-Müllerian hormone produced by Sertoli cells in the testis triggers regression of the Müllerian ducts. The exact chemical structure of the factor responsible for Müllerian duct regression, a glycoprotein of 124,000 MW (30), is not yet known. Müllerian ducts are sensitive to anti-Müllerian hormone only during a short period of development, up to the 8th gestational week in the human fetus. Exposure to fetal testis after this date does not prevent normal Müllerian duct development. Müllerian ducts are not affected by testosterone or its derivatives at any developmental stage (31). When anti-Müllerian hormone is missing or ineffective, a uterus and fallopian tube develop in otherwise normal men (32), this condition is termed "persistent Müllerian duct syndrome". This disorder is often detected because the uterus and fallopian tubes prolapse into an inguinal hernia.

Wolffian ducts, which are initially the excretory ducts of the mesonephros become androgen-dependent only when renal function has been taken over by the definitive kidney (33). When this critical period is reached, Wolffian ducts disappear in the absence of androgens. If androgens are present, the Wolffian ducts become stabilized. Androgen stimulation is needed for their subsequent differentiation into male sex accessory organs. However, Wolffian ducts are not maintained in the human female by systemic testosterone administration, nor are they maintained in female pseudohermaphrodites even when the virilizing agent reaches the fetus early in pregnancy (34). Apparently, the human Wolffian duct is sensitive only to testosterone produced locally by the fetal testis.

Differentiation of urogenital sinus and external genitalia (see Fig. 2)

Until the eighth week the external genitalia of both sexes are identical and are capable of differentiation in either direction (35). The common primordial anlagen are: the genital tubercle, the urethral folds and the labioscrotal swelling. The paired urethral folds surround a urogenital slit. Masculinization begins in the normal fetus by a lengthening of the anogenital distance. The labioscrotal swellings fuse in midline forming the scrotum and the sinus of the urethral folds close forming the primordium of the cavernous urethrea. The genital tubercle becomes the corpora cavernosa and the glans penis. Formation of the penis is complete when the fetus is 12-14 weeks old and labioscrotal fusion cannot be achieved in female fetuses exposed to androgens after this period (35). Growth of the penis continues, however, throughout gestation and is normally mediated by the pituitary dependent Leydig cell stimulation of the fetal testis.

In the eight week old female fetus the genital tubercle becomes the clitoris, the labioscrotal swelling the labia majora and the urethral fold the labia minora. Cavitation for vaginal development is first observed at 15 weeks and is complete at 18 weeks. In males, or testosterone-exposed females, caudal growth of the rudimentary vaginal cord is inhibited and canalization of this rudimentary structure yields the prostatic utricle which opens just beneath the bladder neck, between the orifices of the vasa deferentia (36).

Role of dihydrotestosterone (DHT). Testosterone is the androgen that differentiates the Wolffian duct of the human embryo (27) and the active intracellular androgen binding to the Wolffian duct and androgen receptor. In other androgen dependent areas, notably the urogenital tubercle, testosterone acts only as a prehormone (37). In these tissues, testosterone is metabolized by a steroid 5α-reductase, a microsomal enzyme, to DHT which governs the morphogenetic effects of androgens in these tissues (Fig. 3). In the urogenital tubercle, high activity of the 5α-reductase at the time of sexual differentiation allows testosterone to be metabolized to DHT for preferential binding to the androgen receptor which probably binds both testosterone and DHT. In most tissues, however, the receptor shows a much higher binding affinity for DHT than for testosterone. Once formed the androgen-receptor complex is transferred into the nucleus where it binds to chromatin and initiates gene transcription and increases specific protein synthesis (38, 39). Genes regulating testosterone and DHT receptors are X-linked (40). The reason that testosterone mediates some androgen effects and DHT mediates others is not clear. It may involve subtle differences in the affinity of the receptor for the androgen and/or some aspect of the metabolism of testosterone (8). For example, increase in muscle mass and voice changes in male puberty appear to be mediated by testosterone. Acne, beard growth, temporal hairline recession and prostatic growth appear to be mediated by DHT (41).

Abnormal Male Sexual Differentiation

1. Male pseudohermaphroditism secondary to:

A. Placental and fetal pituitary gonadotropin deficiency. Placental HCG deficiency has not been described. It is conceivable that such a defect can occur; the consequences would be markedly impaired sexual differentiation, since male differentiation until the 20th week is entirely under HCG control. Fetal pituitary FSH and LH are not required for normal genital differentiation but they do play a role in normal genital growth in the 3rd trimester.

Prader et al. (42) invoked prenatal LH deficiency as a

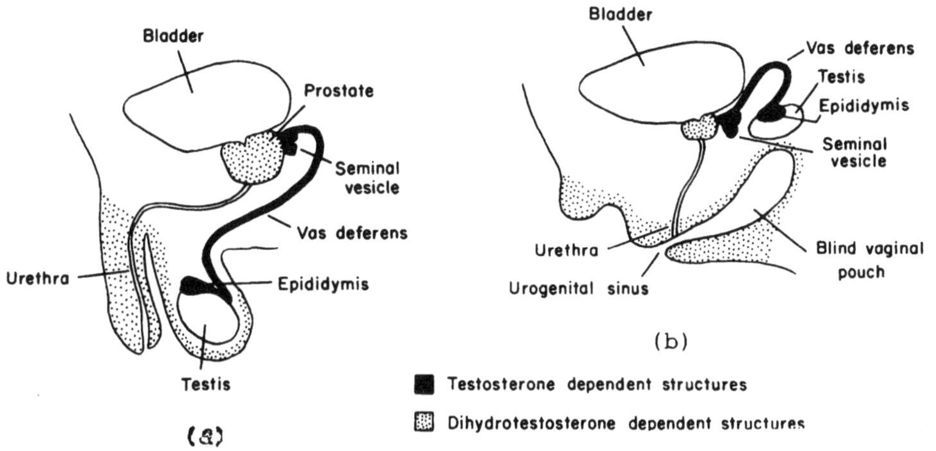

Fig. 3 (a) Suggested selective roles of testosterone (blackened structures) and dihydrotestosterone (stippled structures) in normal male sexual differentiation. (b) Findings in males who cannot convert testosterone to dihydrotestosterone (5α-reductase deficiency): the testosterone dependent structures are normally formed although, as illustrated, testes do not descend. Dihydrotestosterone dependent structures are not developed. A blind ending vaginal pouch is also frequently present (redrawn from 61).

possible cause of MPH, hypospadias, micropenis and cryptorchidism. This postulated pathophysiological mechanism is at variance with the concept that fetal gonadotropins assume control of testicular function only after the period of genital differentiation. Detailed evaluation of pituitary function in these patients, at a later date, would appear to be indicated since deficiency of other pituitary hormones (particularly growth hormone and TSH) has to be ruled out.

B. Leydig cell agenesis. Berthezène et al. (43) described a 35-year-old MPH with a specific defect in testicular Leydig cell differentiation. The patient had a female phenotype with a shallow vagina, intra-abdominal testes and vasa deferentia and an epididymis were present. No Müllerian structures were present. At biopsy the testis contained no recognizable Leydig cells. Serum testosterone was low, LH was elevated, yet FSH was normal. There was no response to HCG stimulation. The testes produced anti-Müllerian hormone normally and testosterone at least during some time in fetal life since Wolffian differentiation was completed.

While these disorders had been attributed to primary Leydig cell agenesis, it was suggested (44) that the findings would best be explained by end organ unresponsiveness to HCG and LH due to a deficiency of HCG-LH receptors and plasma membranes of fetal and postnatal Leydig cells. *In vivo* and *in vitro* studies demonstrated lack of binding of ^{125}I labelled LH to Leydig cells. This form of MPH is also similar to an animal model in which a LH receptor defect is proposed (45, 46).

C. Congenital anorchia (vanishing testes syndrome). Congenital anorchia, or the absence of one or both testes in a 46 XY phenotypically male individual has been estimated to occur in about 1 in 5000 or 1 in 20,000 male births respectively (47). Familial aggregates have been reported (48). According to Jost's work (2) the testes must have been functional until at least the 16th week of gestation in order to produce normal male internal ducts. Müllerian regression and male genitalia develop normally. A testicular vascular accident producing torsion, infarction and loss of one or both testes is the likely explanation for this syndrome. Absence of a testosterone response to HCG stimulation confirms the bilateral anorchia in these patients. Although the patient is similar to a prepubertal castrate, gonadotropins may not be elevated in early childhood (49). If the condition is bilateral, hormonal replacement at puberty is necessary. At puberty, silastic prostheses should be implanted into the scrotum, which is often hypoplastic.

2. Disorders of male sexual differentiation due to abnormal testosterone synthesis

Testosterone is essential for virilization of the male genital tract. Five distinct, autosomal recessive, genetic disorders may result in diminished testosterone synthesis leading to incomplete virilization of the male embryo and MPH. Heterozygous subjects are phenotypically normal. MPH describes an individual with ambiguity of the internal or external genitalia whose karyotype is 46 XY and who has testes. It can be caused by defects in testosterone synthesis, testosterone metabolism (5α-reductase deficiency) and androgen action.

The decrease in testosterone biosynthesis is due to a defect in one of the five enzymes (50) required for the conversion of cholesterol to testosterone (Fig. 4). Deficiency of 20-22 des-

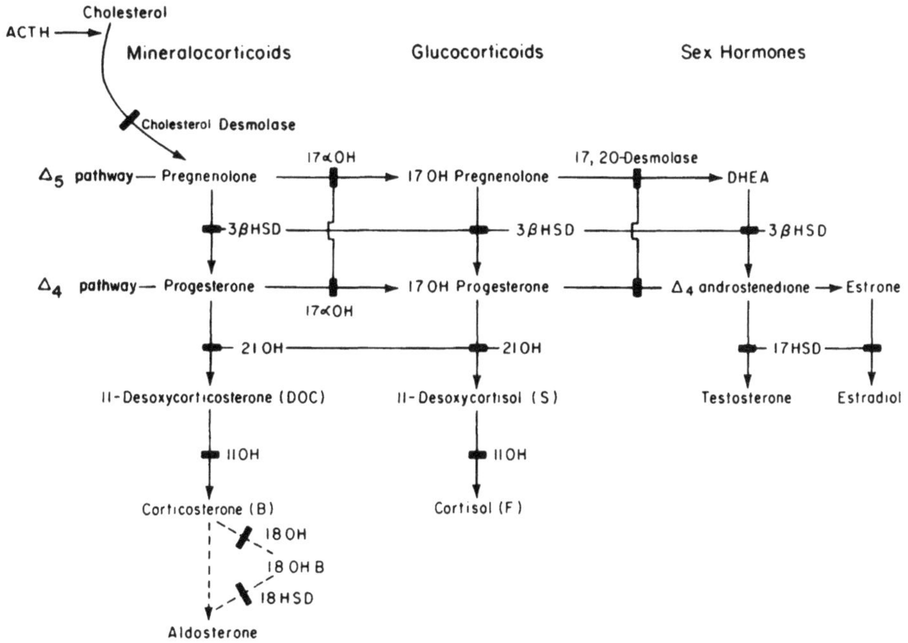

Fig. 4 Biosynthesis of mineralocorticoids, glucocorticoids and sex steroids.

molase, 3β-hydroxysteroid dehydrogenase or 17α-hydroxylase result in cortisol as well as testosterone deficiency and are forms of congenital adrenal hyperplasia. Deficiency of 17-20 desmolase or 17β-hydroxysteroid dehydrogenase affect primarily testosterone biosynthesis by the testes (Fig. 4 and Table 1). Production of testosterone in these forms of MPH is low and that of its precursor is high. Androgen levels in the unstimulated state or after HCG stimulation will often be diagnostic. An HCG test may also be used to measure the ability of testes to produce testosterone in prepubertal children.

Inhibition of the Müllerian ducts is usually complete in these patients, masculinization of the external genitalia is not. This suggests that secretion of anti-Müllerian hormone by the Sertoli cells is independent of testosterone synthesis.

A. Cholesterol desmolase deficiency. This form of male pseudohermaphroditism is characterized by severe salt wasting and impaired synthesis of mineralocorticoids, glucocorticoids and sex steroids. The biosynthetic defect involves the side chain cleavage converting cholesterol to pregnenolone. Conversion of cholesterol to pregnenolone requires a complex mitochondrial mixed function oxidase system including cytochrome P-450. Recently a decrease in the adrenal mitochondrial content of cytochrome P-450 was shown in patients with this defect (51). Plasma steroids and urinary 17-ketosteroid and 17-hydroxysteroid excretion are low and are not stimulated by ACTH. No C_{21}, C_{19} or C_{18} steroids are synthesized by the adrenal glands or the gonads. In a genetic male, with this defect, female external genitalia with male genital ducts and a blind ending vaginal pouch were found at autopsy. The enlarged adrenals were remarkable in that the cortical cells were filled with lipoid material consisting of cholesterol and cholesterol esters, which prompted the name lipoid adrenal hyperplasia (52, 53). The enlarged adrenals can be documented radiographically.

Several patients with this biosynthetic defect have been described. Most died in infancy in adrenal crisis because of

TABLE 1

Clinical and laboratory features of various forms of male pseudohermaphroditism

Clinical features: Newborn with sexual ambiguity – Female	Newborn with sexual ambiguity – Male	Salt wasting	Hyper-tension	Postnatal virilization	Pubertal virilization	Enzyme deficiency	Urine: 17KS	Urine: 17OH	Urine: P'triol	Urine: Aldo	Urine: E/A	Blood: 17OHP	Blood: DHEA	Blood: DS	Blood: Δ4	Blood: T	Blood: DHT	Blood: T/DHT
0	+	+	0	0	0	cholesterol desmolase	↓↓	↓↓	↓↓	↓	?	↓	↓	↓	↓	↓	↓	?
+	+	+	0	0	+	3β HSD	↑*	↓↓	↓	↓	nl	↑	↑↑	↑↑	↓	↓	↓	?
0	+	0	+	0	0	17-hydroxylase	↓↓	↓↓	↓↓	↓	nl	↓	↓	↓	↓	↓	↓	?
0	+	0	0	0	0	17-20 desmolase	↓↓	nl	nl**	?	nl	↑	↓	↓	↓	↓	↓	
0	+	0	0	0	+	17 HSD	↑↑	nl	↑	nl	nl	↑	↑	↑	↑↑	↑	↑	nl
0	+	0	0	0	+	5α-reductase deficiency	nl	nl	nl	nl	↑↑	nl	nl	nl	nl	nl	↓	↑***

* Mostly Δ^5 -17-ketosteroids (e.g. DHEA).

** Pregnanetriolone markedly elevated.

*** Abnormality in infancy; in prepubertal children only after HCG stimulation and adults.

Abbreviations: P'triol, Pregnanetriol; Aldo, Aldosterone; E, Etiocholanolone; A, Androsterone; 17OHP, 17OH Progesterone; DS, Dehydroepiandrosterone sulfate. For other abbreviations, see text.

(Adapted from Saenger P., Levine, L.S., New, M.I. (1980) Sexual ambiguity at birth. In Bain J. Hafez ESE, eds. Diagnosis in andrology. p. 46. The Hague: Martinus Nijhoff).

deficient glucocorticoid and mineralocorticoid production. The severe ambiguity of the external genitalia in the male suggests a defect also in the testes. Affected female infants have normal genital development. Because the defect impairs aldosterone and desoxycorticosterone secretion, salt-wasting is attributable to the lack of mineralocorticoid secretion. Early recognition and proper treatment with gluco- and mineralo-corticoids should permit survival. All affected XY males have been reared as female. They require gonadectomy and estrogen replacement at puberty.

B. 3β-Hydroxysteroid dehydrogenase (3β-HSD) deficiency. An enzymatic deficiency of 3β-HSD, first described by Bongiovanni (54), also results in decreased synthesis of all three classes of adrenal steroids. There is impaired secretion of aldosterone, cortisol and testosterone, leading to MPH and life-threatening salt-wasting in infancy. The enzyme is deficient in both gonadal and adrenal tissue. Females with 3β-HSD defect show only minimal clitoral enlargement, yet no labioscrotal fusion probably due to the extremely high amounts of DHEA which is a weak androgen. The male with this disorder may resemble the virilized female with salt wasting due to 21-hydroxylase deficiency. Although it is secreted in very high amounts, DHEA is too weak an androgen to achieve complete masculine development of the external genitalia. Congenital adrenal hyperplasia due to 3β-HSD deficiency must be suspected in a newborn of either sex with salt-wasting symptoms who has ambiguous genitalia. Some male infants with this defect may only present with mild hypospadias at birth.

The urine of patients with 3β-HSD deficiency predominantly contains steroids with a double bond at the C-5 (Δ^5-steroids). Excessive urinary excretion of DHEA and its precursor pregnenetriol, with markedly elevated DHEA and DHEA-sulfate levels in plasma, are diagnostic of 3β-HSD deficiency. Since stress in the immediate postnatal period may elevate DHEA and DHEA-sulfate levels in infants with normal adrenal function, DHEA and DHEA sulfate reference values for newborns should be consulted before the diagnosis of 3β-HSD is entertained (55).

3β-HSD in testes, adrenal and liver may be under different genetic control and could result in varying degrees of enzymatic deficiency in different organs. Such a hypothesis may explain the advent of male puberty which has been described in some boys with 3β-HSD deficiency (56).

Florid pubertal breast development in this form of male pseudohermaphroditism may be the result of inadequate testosterone production in fetal life, leaving the breast anlagen unsuppressed (56).

Mild forms of this defect are probably due to partial deficiency of 3β-HSD (57). Partial deficiency of 3β-HSD in adrenal and gonad was found in prepubertal patients with acne, male pattern hirsutism and amenorrhea. Clitoromegaly but not genital ambiguity was present. In some patients an intact hepatic 3β-HSD may convert secreted Δ^5-3β-hydroxysteroids into androstenedione and testosterone. 16α-Hydroxylation of Δ^5-3β-hydroxysteroids was also a prominent pathway, suggesting the persistence of fetal pathways of steroid metabolism in some patients. The fetal adrenal cortex is normally 3β-HSD deficient and is known to 16α-hydroxylate pregnenolone. When the disorder is mild, the baseline pattern of blood hormones may not be characteristic and an ACTH stimulation test will reveal the defect resulting in hyper-responsiveness in 17-hydroxypregnenolone and DHEA levels. Subtle defects in 3β-HSD activity may thus mimic the so-called polycystic ovary syndrome and hirsute adult females may have mild 3β-HSD deficiency (58). Glickman et al. (59) found that 17-hydroxypregnenolone levels are more apt in identifying these patients than DHEAS levels. Corticoid treatment restored ovulatory cycles and fertility and in one unreported case in an adult female, pregnancy was successfully achieved (60).

C. <u>17α-Hydroxylase deficiency</u>. A defect in 17α-hydroxylase also results in diminished secretion of all glucocorticoids and sex steroids (61). Cases have been reported in both males and females (62, 63). Patients with a 46 XY karyotype are born with ambiguous genitalia, while females with this defect have hypogonadism presenting at puberty.

Prominent breast development was the only secondary sex characteristic which occurred at puberty in a male patient (61). The decrease in cortisol production resulted in increased ACTH secretion and an increased production of corticosterone and desoxycorticosterone (61). Hypertension, hypokalemia and alkalosis are frequently present in this syndrome and are probably the result of the markedly increased desoxycorticosterone and corticosterone production (61, 64). The increased desoxycorticosterone levels result in sodium retention, decreased plasma renin activity and depressed plasma and aldosterone secretion. Sex steroid secretion is low and secondary sex characteristics fail to develop at puberty. FSH and LH levels are elevated. Urinary 17-ketosteroid excretion is low. Lifelong therapy with glucocorticoids is mandatory in males and females with this defect to maintain normal blood pressure and to correct the hypokalemic alkalosis (61).

D. <u>17,20 Desmolase deficiency</u>. This defect was first described by Zachmann et al. (65), in three patients - two first cousins and a "maternal aunt" - suggesting an autosomal recessive or X-linked mode of inheritance. All three patients had a 46 XY karyotype. These three patients shared a familial form of male pseudohermaphroditism, due to a partial defect in conversion of 17α-hydroxyprogesterone and 17α-hydroxypregnenolone to C_{19} steroids by testes and adrenals. Only sex steroid biosynthesis is affected in this form of male pseudohermaphroditism. The patients presented with ambiguous external genitalia and inguinal or intra-abdominal testes with no spermatogonia. All had third degree hypospadias but normal male ductal differentiation. One of the subjects had a rudimentary uterus and fallopian tubes. Another patient had female external genitalia and a 7cm blind ending vaginal pouch.

Urinary pregnenetriol, a metabolite of 17-hydroxypregnenolone, was increased and increased further after ACTH and HCG stimulation. However, testosterone and DHEA excretion did not rise appreciably. *In vitro* studies with testicular tissue (66) also demonstrated a defect in the conversion of C_{21} steroids to C_{19} steroids. The results were consistent with a deficiency of adrenal and testicular 17,20-desmolase activity. Two types of MPH due to 17,20-desmolase deficiency may exist: one with an incomplete defect in both the Δ^4 and the Δ^5 pathway and one with a more or less complete defect in the Δ^4 pathway only (67).

The diagnosis of 17,20-desmolase deficiency can be established by demonstrating the inability of adrenals and testes to convert C_{21} to C_{19} steroids. Sex steroid therapy is necessary in both sexes at the age of puberty. Gonadectomy is recommended in 46 XY patients raised as females.

E. 17-Hydroxysteroid dehydrogenase - 17-HSD deficiency. Deficiency of the 17-HSD enzyme results in decreased testosterone synthesis and is another cause of MPH. With one exception all previously reported genetic males were assigned female sex at birth because they were so incompletely masculinized (68, 69). One patient who presented with sexual ambiguity noted in infancy (70) was given a male sex assignment.

The hyperplastic Leydig cells in these patients secrete predominantly androstenedione, a testosterone precursor. In the testicular effluent the androstenedione/testosterone ratio was 96:5 (normal 3:75) (68), whereas in peripheral blood it was less elevated (5.8; normal 0.25-1.05) showing that peripheral conversion to testosterone was apparently still intact. This suggests that the enzymes controlling 17-HSD in testes and liver are under different genetic control. Conversion of estrone to estradiol is also impaired but usually to a lesser degree.

FSH may be normal while LH is elevated. LH can be suppressed by testosterone administration. The relative lack of intrauterine virilization may be due to efficient aromatization of androstenedione to estrone in the placenta so that little androstenedione is available for extratesticular conversion to testosterone (71).

At puberty, these patients display heterogeneity in their phenotype. Varying concentrations of circulating estrogens may explain varying degrees of breast development noted in patients with 17-HSD.

The increased amounts of androstenedione and estrone secreted at puberty are converted peripherally to testosterone and estradiol which may explain the simultaneous occurrence of gynecomastia and virilization in some of these patients (72). Normal 5α-reductase activity is present though as a consequence of low testosterone levels DHT is decreased. It has also been shown by Saez et al. (73) that nitrogen retention in response to testosterone is preserved. Despite the fact that external genitalia are only partially virilized, Wollfian ducts are normal.

The defect in androstenedione and estrone metabolism is also demonstrable with testicular incubation studies (68).

Demonstration that androstenedione is not converted to testosterone is necessary for diagnosis in 17-HSD deficiency. Since breasts and virilization develop at puberty, the patients may require treatment to make the puberty conform with the sex of assignment.

3. 5α-Reductase deficiency - A disorder of testosterone metabolism

5α-Reductase deficiency is the first recognized inherited abnormality of steroid metabolism in which the basic defect resides in the hormone responsive target tissues (74). This disorder has fundamental implications for understanding sexual differentiation in man.

5α-Reductase deficiency, an autosomal recessive defect, causes abnormally low conversion of testosterone to DHT, the androgen responsible for masculinizing the external genitalia of the male fetus (Fig. 3). In males 5α-reductase deficiency leads, therefore, to pseudohermaphroditism. Homozygous females with this deficiency have a normal phenotype and are fertile, since DHT does not play a role in female sexual differentiation or pubertal maturation. Heterozygous 46 XY individuals are clinically normal (41).

Boys with 5α-reductase deficiency are born with ambiguous genitalia. There is an enlarged clitoris, often with ample prepuce and a ventral urethral groove with a chordee. A urogenital sinus is present and, depending upon the degree of masculinization, either a single orifice or separate urethral (anterior) and vaginal (posterior) orifices may be present. The labia majora are rugated and the labia majora are sometimes fused. Usually a blind-ending vaginal pouch is present. This blind-ending pouch varies considerably in size: in some patients it is only 0.5 cm deep, in others it may measure 5 cm and it may also be completely absent (75). The pouch can be filled easily during a voiding cystourethrogram. The internal genitalia consist of testes and epididymis which are found in the abdomen, in the inguinal area, or in the labia majora (41).

At puberty, the phallic clitoris enlarges. The testes too, enlarge, and descend into the now rugated labioscrotal folds. A male body habitus develops, the voice deepens, and frequent erections occur (41).

Acne is usually absent in these patients. Adult males also have a decreased beard and no temporal hairline recession. The prostate remains small, and even in males over 30 years of age, it is barely palpable (41). The lack of these features of normal male development suggests that they are DHT dependent. Breast tissue is not palpable. In this form of MPH normal secretion of testosterone probably suppresses breast anlagen *in utero* (71).

The diagnosis of 5α-reductase deficiency can be confirmed by demonstrating normal levels of testosterone before and after HCG administration and a deficient DHT response leading to a markedly increased testosterone/DHT ratio (75).

In a child born with ambiguous genitalia, rapid diagnosis of a steroid enzyme deficiency is particularly desirable. In young male infants testosterone is physiologically elevated to low adult levels for a period of between 10 and 60 days of life (77). Measurement of testosterone and DHT will, therefore, be diagnostic during this time period and may obviate the need for an HCG stimulation test (78). The normal testosterone/DHT ratio in young male infants is 4.9 ± 2.8 (mean ± SD); this ratio does not change significantly after HCG stimulation and is lower than that in older children and adults.

In older infants and in prepubertal males, when testosterone is normally very low, HCG stimulation is necessary to assess the testosterone/DHT ratio. In normal children a DHT rise paralleling the increase of testosterone is expected after a course of HCG injections. The normal ratio in children is 11 ± 4.4 (79). The measurement of testosterone/DHT ratios after the administration of HCG is, therefore, a useful adjunct in diagnosing 5α-reductase deficiency in childhood (79).

In adult patients circulating testosterone is elevated, probably secondary to increased LH levels. The reason for the high LH levels is unclear (41). All patients form some DHT after the time of expected puberty as the enzyme defect is never complete. 5α-Reductase deficiency is also expressed in the urinary metabolites of testosterone. For schematic presentation of the biotransformation of testosterone see Fig. 5. The ratio of the urinary 5α-reduced 17-ketosteroid metabolite etiocholanolone (5β3α) to the 5α-reduced metabolite androsterone (5α3α) is less than 2 in normal children and adults (79). There is no sex difference in the etiocholanolone/androsterone ratio in children (79). In patients with 5α-reductase deficiency the etiocholanolone/androsterone ratio is always above 2. In the larger series of Peterson et al. (41) the etiocholanolone/androsterone ratio in affected individuals was 5.02 ± 2.26. Measurement of urinary etiocholanolone/androsterone ratios is also useful for the detection of obligate heterozygote carriers (41).

In normal subjects the rate of DHT formation from testosterone was greater in fibroblasts grown from prepuce and scrotum

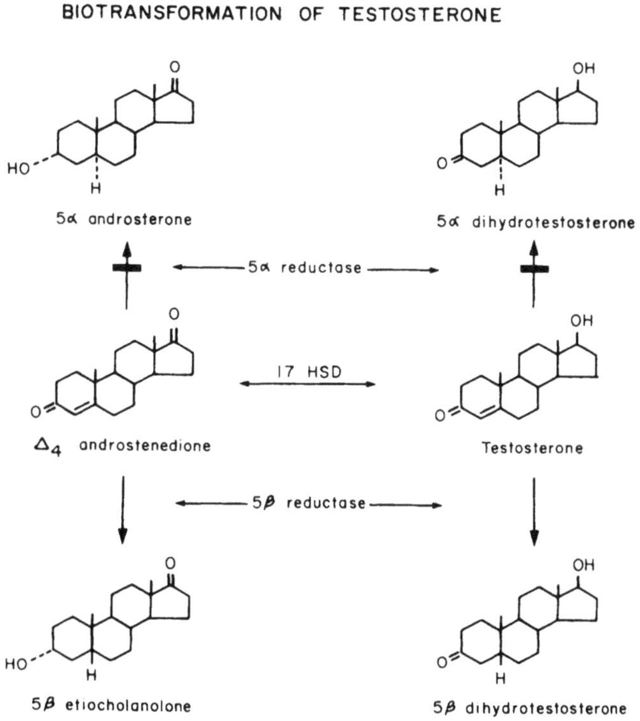

Fig. 5 Biotransformation of testosterone. Black bars indicate enzymatic block in conversion of testosterone and androstenedione to their respective 5α-reduced metabolites.

than in fibroblast from non-genital sites (80). DHT formation is low in fibroblasts from patients with 5α-reductase deficiency. 5α-Reductase activity in cloned normal human skin fibroblasts may however undergo a shift from high to low activity upon sub-cloning (81).

Several distinct mutations may affect *in vitro* 5α-reductase activity. In one family, the mutation probably involved the interaction of enzyme and steroid substrate leading to a decreased conversion of testosterone to DHT. A study of another pedigree indicates the presence of an inherently unstable enzyme in certain *in vitro* conditions such as exposure to cycloheximide or elevated temperature (82). Recently, cells from another patient were found to have intermediate levels of activity and stability of 5α-reductase (83). Griffin and Wilson (39) point out this genetic heterogeneity is akin to that noted in other enzymatic deficiencies, such as glucose-6-phosphate dehydrogenase deficiency (84).

The level of androgen receptors for DHT is normal. The clinical observation that DHT treatment caused enlargement of the prostate, beard growth and acne is in concordance with this biochemical finding (41).

Gender role and gender reversal in 5α-reductase deficiency. The largest pedigree with 5α-reductase deficiency has been found in an isolated village in the Dominican Republic (41). This rare form of MPH afflicts some 1% of all males in this community, causing 38 affected individuals. In the past, 18 had been reared allegedly unambiguously as females. During or after puberty with the development of male secondary sex characteristics including phallic enlargement and increased muscle mass, 16 of 18 changed their assigned sex and sex of rearing from male to female. Sex reversal at the age of puberty has not been reported in the patients studied in the United States. Change of assigned gender role after puberty has been reported sporadically in patients with MPH due to other causes (85).

The findings in the Dominican Republic raise the possibility that postnatal and pubertal androgen action contribute towards the development of male gender identity (86). If this were true, it would certainly be at variance with the widely held concept that psychosexual orientation and gender awareness in man is irrevocably determined by the sex of rearing in the first 5 years of life and that gender identity follows the sex of rearing (87). The pressures to conform to the sex of rearing in a more modern society are very great, yet they may not be applicable to children with 5α-reductase deficiency reared in rural Santo Domingo (88). Considering the fact that they were known in their community as "guevedoces" (penis at 12 [years]), it seems likely that these individuals were not unambiguously raised as females, but rather had a specific status as hermaphrodites with the portentous knowledge of an impending sex change at puberty (88).

Nonetheless, it is possible that under certain conditions, the cumulative effects of androgen exposure of the brain *in utero*, in the early postnatal period and again at puberty may predominate and may override the effects of rearing practices as girls in infancy and childhood.

The case studies from the Dominican Republic thus provide impressive circumstantial evidence that both hormone and psychosocial factors influence gender identity and gender role in man (1). Wilson (89) concluded however that it is difficult to design rigorous scientific studies in humans to determine the mechanisms by which the two types of determinants interact.

<u>Sex assignment in MPH due to 5α-reductase deficiency</u>. When diagnosed in infancy, the choice of gender should be based on the infant's phenotype. If appropriate plastic repair of the external genitalia is possible, these infants should be reared as males. If the phallus remains very small, even after androgen therapy, surgical correction of perineal hypospadias is almost impossible, and the final result often less than satisfactory. It is then wiser and more gratifying to rear a genetic male with this type of MPH as a female, for it is relatively easy for the surgeon to create a vagina. Castration should be performed well before puberty to avoid disturbing virilization. Older children with this syndrome, in whom gender identity is female, should undergo castration before puberty, clitoroplasty and estrogen therapy at the age of puberty.

Female Pseudohermaphroditism (FPH) from Fetal Testosterone Overproduction due to Defects of Glucocorticoids and Mineralocorticoid Synthesis

A female fetus will become masculinized if exposed to excessive androgens *in utero*. The degree of masculinization is dependent upon the time of exposure to excessive androgens. If exposed after the 12th week when the vagina has already been separated from the urethra only clitoral hypertrophy will occur. If before that date, masculinization may be so profound that the urethral opening is "penile".

Congenital adrenal hyperplasia is the most common cause of FPH. Since the female fetus does not possess a testis, anti-Müllerian hormone is not produced and the female with congenital adrenal hyperplasia is born with a normal uterus and fallopian tubes. Female genital abnormalities are present only in the androgen responsive external genitalia. The usually complete involution of the Wolffian system, despite increased fetal androgen, suggests that the level of androgen necessary for Wolffian development is higher than that produced in congenital adrenal hyperplasia. It is possible that only testicular androgens achieve high enough concentrations necessary for Wolffian duct stabilization.

Three forms of congenital adrenal hyperplasia, 3β-HSD deficiency, 21-hydroxylase deficiency and 11-hydroxylase deficiency, cause FPH. They can readily be diagnosed by measurements of 17-hydroxyprogesterone, DHEA, and 11-desoxycortisol and desoxycorticosterone respectively (see Fig. 5). For an excellent update on congenital adrenal hyperplasia, see reference 90.

Effect of maternal ingestion of androgens and progestins

FPH has been observed after maternal ingestion of androgens (testosterone and alkylated derivatives), progestins (progesterone and 17α-hydroxyprogesterone), synthetic progestins (medroxyprogesterone, norethindrone, ethisterone, norethynodrel), and aminoglutethimide (91). There is a low risk of fetal virilization from oral contraceptives which contain small amounts of progestins. To cause formation of a urogenital sinus these drugs have to be ingested before the 12th week of pregnancy; after that they may only cause clitoral enlargement.

Disorders of Androgen Action

a. Testicular feminization

In the syndrome of testicular feminization the karyotype is 46 XY. Under the influence of the Y chromosome, the gonads differentiate as testes. The testes secrete testosterone and DHT, but because of a receptor defect caused by mutation of a gene on the X chromosome (40), there is no response to testosterone or DHT and these individuals develop as phenotypic females. The fetal testis of these females secrete anti-Müllerian hormone normally. Development of the fallopian tubes, uterus and the upper third of the vagina is thus suppressed (92).

Patients with testicular feminization are detected at birth, because they have an inguinal hernia or after puberty, because of primary amenorrhea. At puberty, affected individuals show normal breast development. External genitalia are unambiguously female and the clitoris is normal in size. The vagina ends blindly and may be shortened depending on the relative contribution of Müllerian ducts and the urogenital sinus to the formation of the vagina (92). About a third of the patients have negative family histories and are presumed to represent new mutations. Testicular biology is similar to that in long-standing cryptorchidism; spermatogenesis is absent.

The incomplete disorder resembles the complete form except that there is genital ambiguity and virilization as well as feminization may occur at puberty (39, 92). The risk of testicular neoplasis is small in the first two decades. Prior to age 15, the risk of neoplasis is 2.6%. Thereafter the risk increases to 20-30% by age 40-50 years (93). Therefore, testes should remain *in situ* until pubertal growth and feminization is completed and be promptly removed thereafter.

Androgen production is normal or higher than normal (39). The elevated production rate is probably secondary to a high LH level which in turn may be the resultant of deficient feedback regulation caused by androgen resistance at the hypothalamic-pituitary level. Higher LH levels lead to increased estrogen production by testes contributing to the appearance of normal female secondary sex characteristics at puberty. The androgen resistance in this syndrome is due to abnormalities of the androgen receptor. It has now been demonstrated that there are at least two genetic variants: the first is characterized by undetectable DHT binding to the cytosol of fibroblasts. This is due either to the absence of the receptor protein (receptor negative form of testicular feminization) or alternatively due to changes in the receptor structure resulting in a temperature dependent loss of affinity for the steroid (receptor negative form of testicular feminization). Unstable receptors have been found in both complete and incomplete testicular feminization. The second variant demonstrates normal cytoplasmic as well as nuclear binding of DHT (receptor positive form of testicular feminization). These patients may represent a defect residing at some site distal from the receptor, such as the site of generation of specific messenger RNA in the nucleus (94). (See Fig. 6).

It is important for the clinician to recognize these two variants, particularly in patients with incomplete testicular feminization. An XY individual with inguinal testes and complete external genitalia with undetectable androgen binding may be expected to feminize spontaneously at puberty. However, an infant with normal androgen binding may not be completely insensitive to androgen and may, therefore, virilize at puberty (96). Presently there is no reliable and simple way to predict the response of cells in these patients with conserved androgen binding. Androgen response to HCG stimulation and/or effects of

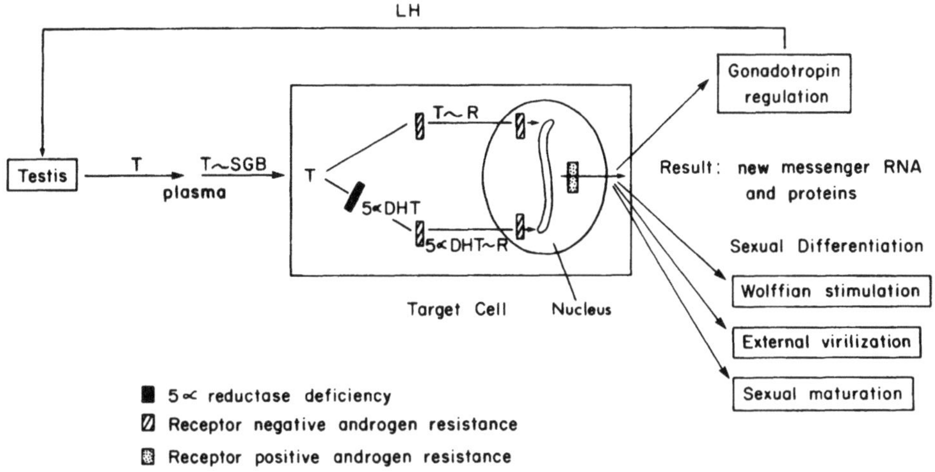

Fig. 6 Mechanisms by which androgens virilize the male embryo. Three types of mutations in this pathway have been informative in proving the applicability of this model to the events in the male embryo: those that result in abnormalities in the 5α-reductase enzyme and disorders of the androgen receptor, i.e. receptor negative resistance and receptor positive androgen resistance; T, testosterone; DHT, dihydrotestosterone; SGB, sex hormone binding globulin; R, high affinity androgen receptor protein. (Adapted from 39).

testosterone on phallic size may be of help in arriving at a final decision, although these diagnostic measures may lead to an undesirable delay in sex assignment (97).

In normal males an increase in serum testosterone leads to a lowering of the androgen binding protein. Recently the response of androgen binding protein to HCG stimulated testosterone has been suggested as a useful tool to determine post-receptor response in receptor positive patients with MPH or incomplete testicular feminization (98).

b. Reifenstein's syndrome

In addition to incomplete testicular feminization, Reifenstein's syndrome may present as MPH (99).

Most commonly one observes a male neonate with perineoscrotal hypospadias, who will develop azoospermia and gynecomastia after puberty. Cryptorchidism is common and the testes are small. Androgen receptor levels in fibroblasts from these patients are partially deficient. The receptors are not thermolabile. Hypospadias, cryptorchidism and gynecomastia require surgical correction.

Despite the lack of virilization at puberty these subjects have levels of plasma androgens which are within or slightly above the normal adult male range. Levels of plasma DHT are normal. LH levels are uniformly elevated.

It should be noted that the familial syndrome of MPH (100, 101, 102) probably represents phenotypic variations of partial Reifenstein's syndrome.

The disorder is inherited as an X-linked recessive or autosomal dominant male-limited tract.

Management of Patients with Ambiguous Genitalia

The birth of an infant with sexual ambiguity is a medical emergency so that rational sex assignment can be achieved within a few weeks. The external genitalia will rarely be distinctive enough to allow diagnosis of a particular disorder or to distinguish clearly between male and female pseudohermaphroditism. Some affected male infants develop a female phenotype with complete failure of virilization of the Wolffian ducts, urogenital sinus and external genitalia. At the other end of the spectrum, patients with MPH appear as entirely normal males except for e.g. first degree hypospadias. Indeed, MPH may result in a phenotype indistinguishable from that seen in female pseudohermaphrodites. This potential source of grave error in diagnosis

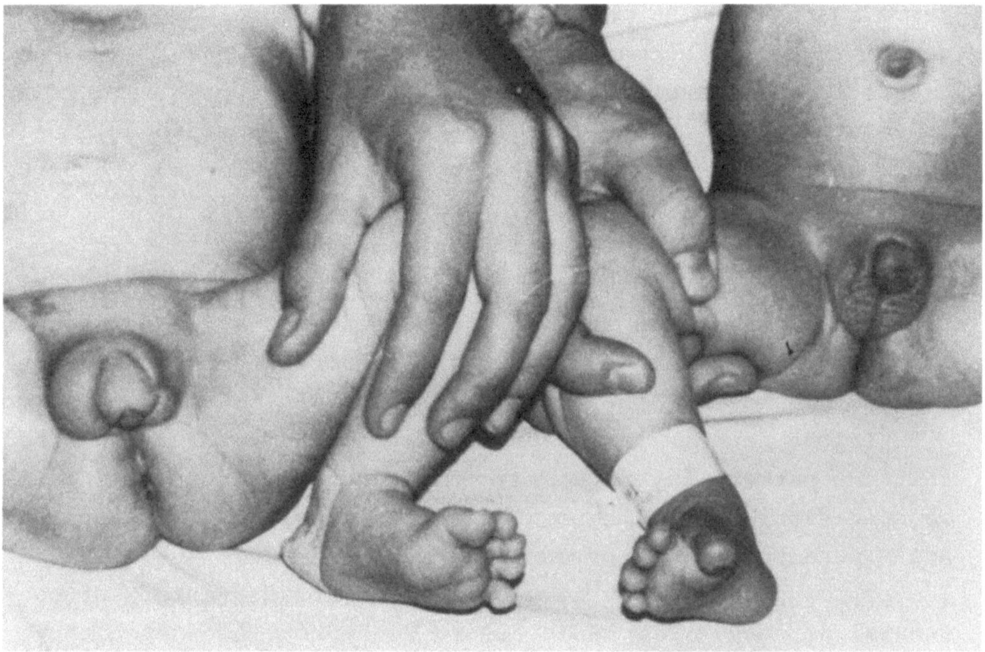

Fig. 7 Similarity in phenotype of genitalia of two infants. Infant on left has male pseudohermaphroditism of undetermined etiology. Infant on right has female pseudohermaphroditism due to 21-hydroxylase deficiency, salt-wasting variety. This picture illustrates that it is not possible to classify patients with ambiguous genitalia on the basis of external genitalia. (Courtesy of Dr. Robert McVie).

and subsequent management emphasizes the need for biochemical characterization of the enzymatic defect in every single patient with MPH.

Palpable gonadal structures contain almost always testicular tissue. Careful rectal examination, radiographic studies and/or abdominal sonography will demonstrate presence or absence of Müllerian derivatives. Presence of Müllerian ducts is evidence against MPH as they are only seen in female pseudohermaphroditism, true hermaphroditism and forms of gonadal dysgenesis.

A buccal smear and chromosomal analysis should be initiated immediately. H-Y antigen serology still requires prohibitively large amounts of blood to be of help in the immediate newborn period.

Urinary steroids as well as plasma androgens before and after ACTH and HCG should be determined in patients with MPH in order to ascertain whether the patient has a block in testosterone synthesis or metabolism (5α-reductase deficiency) involving the testis, the adrenal or both.

In the newborn period the pediatrician must bear in mind that the patient may be at risk for adrenal crisis until forms of congenital adrenal hyperplasia with impaired glucocorticoid and mineralocorticoid production have been ruled out. Elevated 17α-hydroxyprogesterone levels in serum are diagnostic of congenital adrenal hyperplasia due to 21-hydroxylase deficiency (102). Particularly careful attention has to be paid to potential electrolyte imbalance in patients with salt-losing congenital adrenal hyperplasia.

The issue to gender assignment should consider the infant's capacity for future function as a male. The serum androgen response to HCG stimulation and/or effects of systemic testosterone on penile size (97) may be of help in arriving at a final decision although these diagnostic measures may delay sex assignment.

The sex assignment in infants with MPH should be guided by the anatomy of the external and internal genitalia, i.e. the degree of masculinization, the possibility that puberty will conform to the assigned sex and the capacity for normal sexual activity. Except in testicular feminization, gonadectomy prior to puberty is always recommended in 46 XY patients raised as females. Sex steroid therapy concordant with the sex of rearing at the age of puberty is necessary. If reconstructive surgery of the external genitalia is necessary, it should be initiated sometime before the age of 3 years, before the child's gender awareness is embedded too firmly. Money (104) has suggested a parallel between the process of gender identity formation and the acquisition of one's native language. Certainly by the age of five at the latest, children grasp linguistic principles that are involved in the use of their native tongues. Analogously, at this same age the primary identification with one sex or the other is irrevocably part of the self image (105).

If the micropenis cannot be corrected hormonally or surgically, then the construction of a vagina is generally easier than the difficult correction of perineal hypospadias in a child with microphallus. At such a time pelvic exploration to determine the structure of internal genital ducts and gonads should also be planned.

It needs to be stressed that even with the most sophisticated diagnostic techniques available, the pathogenesis of MPH remains obscure in many patients (106, 107).

Recalling the complex time schedule of genital differentiation in man, it becomes clear that only transient disturbances in any of the processes involved may alter the male phenotype.

Testosterone synthesis and metabolism, as well as androgen binding, can be readily assessed but testing for androgen action after receptor binding (receptor positive androgen resistance) is not possible at present.

No reliable prediction is possible as to whether an infant with partial androgen resistance will be able to respond to endogenous or exogenous androgens at puberty. This current limitation should be considered when arriving at a choice of sex assignment in an infant with genital ambiguity.

References

1. Rubin, R.T., Reinisch, J.M., Haskett, R.F.: Postnatal gonadal steroid effects on human behavior. Science 211, 1318-1324 (1981).

2. Jost, A., Vigier, B., Perchellet, J.P.: Studies on sex differentiation in mammals. Recent Prog. Horm. Res. 29, 1-41 (1973).

3. Jirasek, J.E.: Morphogenesis of the genital system in the human. In: "Morphogenesis and Malformation of the Genital System, Birth Defects", vol. XIII, No. 2, Eds. Blandau, J. J., Bergsma, D., Stratton Corporation, New York, pp. 13-39 (1977).

4. Jirasek, J.E.: Principles of reproductive endocrinology. In: "Disorders of Sexual Differentiation", Ed. Simpson, J.L., Academic Press, New York, pp. 52-111 (1976).

5. Ohno, S.: Major sex-determining genes. Monographs on Endocrinology, vol. 11, Springer Verlag, New York, Heidelberg, Berlin, pp. 1-140 (1978).

6. Tran, D., Meusey-Dessolle, N., Josso, N.: Anti-Müllerian hormone is a functional marker of foetal Sertoli cells. Nature (Lond.) 296, 411-412 (1977).

7. Pelliniemi, L.J., Niemi, M.: Fine structure of the human foetal testis. I. The interstitial tissue. Z. Zellforsch. mikrosk. Anat. 99, 507-522 (1969).

8. Wilson, J.D., Griffin, J.E., Leshin, M., George, F.W.: Role of gonadal hormones in development of sexual phenotypes. Hum. Genet. 58, 78-84 (1981).

9. Baker, T.G.: A quantitative and cytological study of germ cells in human ovaries. Proc. Rec. Soc. London (Biol.) 158, 417 (1963).

10. George, F.W., Simpson, E.R., Milewich, L., Wilson, J.D.: Studies on the regulation of the onset of steroid hormone biosynthesis in fetal rabbit gonads. Endocrinology 105, 1100-1106 (1979).

11. Lovinger, R.D., Kaplan, S.L., Grumbach, M.M.: Congenital hypopituitarism associated with neonatal hypoglycemia and microphallus. J. Pediat. 87, 1171-1181 (1975).

12. George, F.W., Wilson, J.D.: Endocrine differentiation of the foetal rabbit ovary in culture. Nature (Lond.) 283, 861-863 (1980).

13. Ohno, S.: The role of H-Y antigen in primary sex determination. J. Amer. med. Ass. 3, 217-220 (1978).

14. Goldberg, E.H., Boyse, E.A., Bennett, D., Scheid, M., Carswell, E.A.: Serological demonstration of H-Y (male) antigen on mouse sperm. Nature (Lond.) 232, 478-480 (1971).

15. Haseltine, F.P., Ohno, S.: Mechanisms of gonadal differentiation. Science 211, 1272-1278 (1981).

16. Koo, G.C., Tada, N., Chaganti, R., Hammerling, U.: Application of monoclonal anti H-Y antibody for human H-Y typing. Hum. Genet. 57, 64-67 (1981).

17. Federman, D.D.: The requirements for sexual reproduction. Hum. Genet. 58, 3-5 (1981).

18. Wachtel, S.S., Koo, G.C., Boyse, E.A.: Evolutionary conservation of H-Y (Male) antigen. Nature (Lond.) 754, 270-272 (1975).

19. Krco, C.J., Goldberg, E.H.: H-Y male antigen detection of eight-cell mouse embryos. Science 193, 1134-1135 (1976).

20. Wachtel, S.S., Ohno, S.: The immunogenetics of sexual development. In: "Progress in Medical Genetics", vol. 3, Eds. Steinberg, A.G., Bearn, A.G., Motulsky, A.G., Childs, B., W.B. Saunders Co., Philadelphia, pp. 109-142 (1979).

21. Wachtel, S.S., Koo, G.C., Breg, W.R., Elias, S., Boyse, E.A., Miller, O.J.: Expression of H-Y antigen in human males with two Y-chromosomes. New Engl. J. Med. 293, 1070-1072 (1975).

22. Wachtel, S.S.: The genetics of intersexuality: Clinical and theoretic perspectives. Obstet. and Gynec. 54, 671-685 (1979).

23. Zenzes, M.T., Miller, U., Aschmoneit, I., Wolf, U.: Studies on H-Y antigen in different cell fractions of the testis during pubescence. Hum. Genet. 45, 297-303 (1978).

24. Nagai, Y., Ciccarese, S., Ohno, S.: The identification of human H-Y antigen and testicular transformation induced by its interaction with the receptor site of bovine fetal ovarian cells. Differentiation 13, 155-164 (1979).

25. Ohno, S., Christian, L.C., Wachtel, S.S., Koo, G.C.: Hormone-like role of H-Y antigen in bovine freemartin gonad. Nature (Lond.) 261, 597-599 (1976).

26. Clements, J.A., Reyes, R.I., Winter, J.S.D., Faiman, C.: Studies on human sexual development. III. Fetal pituitary and serum and amniotic fluid concentrations of LH, HCG, FSH. J. clin. Endocrinol. Metab. 42, 9-19 (1976).

27. Wilson, J.D., Goldstein, J.L.: Classification of hereditary disorders of sexual development. In: "Genetic Forms of Hypogonadism, Birth Defects", Orig. Art. Ser. vol. II, Ed. Bergsma, D., Stratton Corporation, New York, pp. 1-16 (1975).

28. Liao, S., Liang, T., Tymozko, J.L.: Structural recognitions in the interactions of androgens and receptor proteins and in their association with nuclear receptor components.J. Steroid Biochem. 3, 401-408 (1972).

29. Cunha, G.R., Chung, L.W.K., Shammon, J.M., Reese, B.A.: Stromal-epithelial interactions in sex differentiation. Biol. Reprod. 27, 19-42 (1980).

30. Josso, N., Picard, J.Y., Tran, D.: A new testicular glycoprotein: anti-Müllerian hormone. In: "Testicular Development, Structure and Function", Eds. Steinberger, A., Steinberger, E., Raven Press, New York, pp. 21-31 (1980).

31. Josso, N., Picard, J.Y., Tran, D.: The anti-Müllerian hormone. Recent Prog. Horm. Res. 33, 117-167 (1976).

32. Sloan, W.R., Walsh, P.C.: Familial persistent Müllerian duct syndrome. J. Urol. 115, 459-461 (1976).

33. Price, D., Zaaijer, J.J.P., Ortiz, E., Brinkmann, A.O.: Current views on embryonic sex differentiation in reptiles, birds and mammals. Amer. Zool. Suppl. 1, 173 (1975).

34. Mürset, G., Zachmann, M., Prader, A., Fisher, J., Labhart, A.: Male external genitalia of a girl caused by a virilizing adrenal tumor in the mother. Case report and steroid studies. Acta Endocrinol. 65, 627-638 (1970).

35. Grumbach, M.M., Ducharme, J.R.: The effects of androgens on fetal sexual development, androgen induced female pseudohermaphroditism. Fertil. Steril. 11, 157 (1960).

36. O'Rahilly, R.: The development of the vagina in the human. In: "Morphogenesis and Malformation of the Genital System, Birth Defects", Eds. Blandau, J.J., Bergsma, D., Stratton Corporation, New York, vol. 13, pp. 123-136 (1977).

37. Imperato-McGinley, J., Peterson, R.E.: Male pseudohermaphroditism: The complexities of male phenotypic development. Am. J. Med. 61, 251-272 (1976).

38. Mainwaring, W.I.P.: The mechanism of action of androgens. Monographs on Endocrinology, vol. 10, Springer Verlag, New York, Heidelberg, Berlin, pp. 1-178 (1977).

39. Griffin, J.E., Wilson, J.D.: The syndromes of androgen resistance. New Engl. J. Med. 302, 198-209 (1980).

40. Migeon, B., Brown, T.R., Axelman, Y., Migeon, C.J.: Studies of the locus for androgen receptor: localization on the human X chromosome and evidence for homology with the Tfm locus in the mouse. Proc. natn. Acad. Sci. U.S.A. 78, 6339-6343 (1981).

41. Peterson, R.E., Imperato-McGinley, J., Gautier, T., Sturla, E.: Male pseudohermaphroditism due to steroid 5α-reductase deficiency. Am. J. Med. 62, 170-191 (1977).

42. Prader, A., Illig, R., Zachmann, M.: Prenatal LH deficiency as a possible cause of male pseudohermaphroditism, hypospadias, hypogenitalism and cryptorchidism. Pediat. Res. 10, 883 (1976).

43. Berthezène, F., Forest, M.G., Grimand, J.A., Claustrat, B., Mornes, R.: Leydig cell agenesis: a cause of male pseudohermaphroditism. New Engl. J. Med. 295, 969-972 (1976).

44. Grumbach, M.M., Conte, F.A.: Disorders of sex differentiation. In: "Textbook of Endocrinology", Ed. Williams, R.H., W.B. Saunders Co., Philadelphia, p. 423 (1981).

45. Bardin, C.W., Bullock, L.P., Sherins, R.J., Mowszowicz, I., Blackburn, W.R.: Androgen metabolism and mechanism of action in male pseudohermaphroditism. A study of testicular feminization. Recent Prog. Horm. Res. 29, 65 (1973).

46. Rogers, R.M., Garcia, A., Van Den Berg, L., Petrik, P.K.F., Snihurowych, W.M., Crockford, P.M.: Leydig cell hypogenesis: A rare cause of male pseudohermaphroditism and a pathological model for the understanding of normal sexual differentiation. J. Urol. 128, 1325-1329 (1982).

47. Goldberg, L.M., Skaist, L.B., Morrow, J.W.: Congenital absence of testes: anorchism and monorchism. J. Urol. III, 840-845 (1974).

48. Hall, J.G., Morgan, A., Blizzard, R.M.: Familial congenital anorchia. Birth Defects 11, 115-119 (1975).

49. Winter, J.S.D., Faiman, C.: Serum gonadotropin concentrations in agonadal children and adults. J. clin. Endocrinol. Metab. 35, 561-564 (1972).

50. Bongiovanni, A.M.: Congenital adrenal hyperplasia and related conditions. In: "The Metabolic Basis of Inherited Disease", Eds. Stanbury, J.B., Wyngarden, J.B., Fredericksen, D.S., McGraw Hill, New York, pp. 868-893 (1978).

51. Koizumi, S., Kyoya, S., Miyawaki, T.: Cholesterol side-chain cleavage enzyme activity and cytochrome P-450 content in adrenal mitochondria of a patient with congenital lipoid adrenal hyperplasia. Clin. chim. Acta 77, 301-306 (1977).

52. Kirkland, R.T., Kirkland, J.L., Johnson, C., Horning, M., Librik, L., Clayton, G.W.: Congenital lipoid adrenal hyperplasia in an eight-year-old phenotypic female. J. clin. Endocrinol. Metab. 36, 488-496 (1973).

53. Camacho, A.M., Kowarski, A., Migeon, C.G., Brough, A.J.: Congenital adrenal hyperplasia due to a deficiency of one of the enzymes involved in the biosynthesis of pregnenolone. J. clin. Endocrinol. Metab. 41, 153-161 (1968).

54. Bongiovanni, A.M.: The adrenogenital syndrome with deficiency of 3β-hydroxysteroid dehydrogenase. J. clin. Invest. 41, 2086-2092 (1962).

55. Grueters, A., Korth-Schutz, S.: Longitudinal study of plasma dehydroepiandrosterone sulfate in preterm and term infants. J. clin. Endocrinol. Metab. 55, 314-320 (1982).

56. Schneider, G., Genel, M., Bongiovanni, A.M., Goldman, A.S., Rosenfeld, R.L.: Persistent testicular Δ^5 isomerase-3β-hydroxysteroid dehydrogenase deficiency in the Δ^5-3β-HSD form of congenital adrenal hyperplasia. J. clin. Invest. 55, 681-690 (1975).

57. Kenny, F.M., Reynolds, J.W., Green, O.C.: Partial 3-beta-hydroxysteroid dehydrogenase and 21-hydroxylase deficiency in a family with congenital adrenal hyperplasia. Pediatrics 48, 756-765 (1971).

58. Bongiovanni, A.M.: Acquired adrenal hyperplasia. Fertil. Steril. 35, 599-608 (1981).

59. Glickman, S.P., Rosenfeld, R.L., Rich, B.H., Lucky, A.W., Hatch, R.: Dexamethasone suppression and dehydroepiandrosterone-sulfate as markers for adult onset adrenogenital syndrome in hirsute women. Clin. Res. 28, 720A (1980).

60. Saenger, P.: Unpublished observations.

61. New, M.I.: Male pseudohermaphroditism due to 17α-hydroxylase deficiency. J. clin. Invest. 49, 1930-1941 (1970).

62. Biglieri, E.G., Herran, M.A., Brust, N.: 17-Hydroxylation deficiency in man. J. clin. Invest. 45, 1946-1954 (1966).

63. Mallin, S.R.: Congenital adrenal hyperplasia secondary to 17-hydroxylase deficiency: two sisters with amenorrhea, hypokalemia, hypertension and cystic ovaries. Ann. intern. Med. 70, 69-75 (1969).

64. Tourniaire, J., Audi-Parera, L., Loras, B., Blum, J., Castelnovo, P., Forest, M.G.: Male pseudohermaphroditism with hypertension due to a 17α-hydroxylase deficiency. Clin. Endocr. 5, 53-61 (1976).

65. Zachmann, M., Völlmin, J.A., Hamilton, W., Prader, A.: Steroid 17,20-desmolase deficiency: a new cause of male pseudohermaphroditism. Clin. Endocr. 1, 369-385 (1972).

66. Goebelsmann, U., Zachmann, M., Davajan, V., Israel, R., Mestman, J.H., Mishell, D.R.: Male pseudohermaphroditism consistent with 17,20-desmolase deficiency. Gynecol. Invest. 7, 138-156 (1976).

67. Zachmann, M., Werder, E.A., Prader, A.: Two types of male pseudohermaphroditism due to 17,20-desmolase deficiency. J. clin. Endocrinol. Metab. 55, 487-490 (1982).

68. Virdis, R., Saenger, P., Senior, B., New, M.I.: Endocrine studies in a pubertal male pseudohermaphrodite with 17-ketosteroid reductase deficiency. Acta Endocrinol. 87, 212-224 (1978).

69. Givens, J.R., Wiser, M.L., Summitt, R.L., Kerber, K.J., Anderson, R.N., Pittaway, D.E., Fish, S.A.: Familial male pseudohermaphroditism without gynecomastia due to deficient testicular 17-ketosteroid reductase deficiency. New Engl. J. Med. 291, 938-944 (1974).

70. Knorr, D., Bidlingmaier, F., Engelhardt, D.: Reifenstein's syndrome, a 17β-hydroxysteroid-oxidoreductase deficiency. Acta Endocrinol. Suppl. 173, 37 (1973).

71. Goebelsmann, U., Horton, R., Mestman, J.H., Arce, J.J., Nagata, Y., Nakamura, R.M., Thorneycroft, I.H., Mishell, D.R.: Male pseudohermaphroditism due to testicular 17α-hydroxysteroid dehydrogenase deficiency. J. clin. Endocrinol. Metab. 36, 867-879 (1973).

72. Goebelsman, U., Hall, T.D., Paul, W.L., Stanczyk, F.Z.: *In vitro* steroid metabolic studies in testicular 17β-reduction deficiency. J. clin. Endocrinol. Metab. 41, 1136-1143 (1975).

73. Saez, J.M., Morera, A.M., de Peretti, E., Bertrand, J.: Further *in vivo* studies in male pseudohermaphroditism with gynecomastia due to a testicular 17-ketosteroid reductase defect. J. clin. Endocrinol. Metab. 24, 598-600 (1972).

74. Imperato-McGinley, J., Guerrero, L., Gautier, T., Peterson, R.E.: Steroid 5α-reductase deficiency in man: an inherited form of male pseudohermaphroditism. Science 186, 1213-1215 (1974).

75. Saenger, P.: Steroid 5α-reductase deficiency. In: "The Intersex Child", Ed. Josso, N., Karger Publications, Basel, pp. 156-170 (1981).

76. Goldman, A.S., Shapiro, B., Neumann, F.: Role of testosterone and its metabolites in the differentiation of the mammary gland in rats. Endocrinology 99, 1490-1495 (1976).

77. Forest, M.G., Sizonenko, P.C., Cathiard, A.M., Bertrand, J.: Hypophysogonadal function in humans during the first year of life. J. clin. Invest. 53, 819-828 (1974).

78. Pang, S., Levine, L.S., Chow, D., Sagiani, F., Saenger, P., New, M.I.: Dihydrotestosterone and its relationship to testosterone in infancy and childhood. J. clin. Endocrinol. Metab. 48, 821-826 (1979).

79. Saenger, P., Goldman, A.S., Levine, L.S., Korth-Schutz, S., Muecke, F.C., Katsumata, M., Doberne, Y., New, M.I.: Prepubertal diagnosis of steroid 5α-reductase deficiency. J. clin. Endocrinol. Metab. 46, 627-634 (1978).

80. Wilson, J.D.: Dihydrotestosterone formation in cultured human fibroblasts. J. biol. Chem. 250, 3498-3504 (1975).

81. Griffin, J.E., Allman, D.R., Durrant, J.L., Wilson, J.D.: Variation in steroid 5α-reductase activity in cloned human skin fibroblasts. J. biol. Chem. 256, 3662-3666 (1981).

82. Leshin, M., Griffin, J.E., Wilson, J.D.: Hereditary male pseudohermaphroditism associated with an unstable form of 5α-reductase. J. clin. Invest. 62, 685-691 (1978).

83. Imperato-McGinley, J., Peterson, R.E., Leshin, M., Griffin, J.E., Cooper, G., Dragli, S., Berenyi, M., Wilson, J.C.: Steroid 5α-reductase deficiency in a 65-year-old male pseudohermaphrodite. J. clin. Endocrinol. Metab. 50, 15-22 (1980).

84. Beutler, E., Yoshida, A.: Human glucose-6-phosphate dehydrogenase variants. Ann. hum. Genet. 37, 151-155 (1973).

85. Imperato-McGinley, J., Peterson, R.E., Stoller, R., Goodwin, W.E.: Male pseudohermaphroditism secondary to 17α-hydroxysteroid dehydrogenase deficiency. J. clin. Endocrinol. Metab. 49, 391-395 (1979).

86. Imperato-McGinley, J., Peterson, R.E., Gautier, T., Sturla, E.: Androgens and the evolution of male-gender identity

among male pseudohermaphrodites with 5α-reductase deficiency. New Engl. J. Med. 300, 1233-1237 (1979).

87. Baker, S.W.: Psychosexual differentiation in the human. Biology of Reproduction 22, 61-72 (1980).

88. Ehrhardt, A.A., Meyer-Bahlberg, H.F.L.: Prenatal sex hormones and the developing brain. Ann. Rev. Med. 30, 417-430 (1979).

89. Wilson, J.D.: Sex hormones and sexual behavior. New Engl. J. Med. 300, 1269-1270 (1979).

90. New, M.I., Dupont, B., Pang, S., Pollack, M., Levine, L.S.: An update of congenital adrenal hyperplasia. Recent Prog. Horm. Res. 37, 105-181 (1981).

91. Simpson, J.L.: Disorders of sexual differentiation: Etiology and clinical delineation. Academic Press, New York, pp. 183 (1976).

92. Migeon, C.J., Brown, T.R., Fichman, K.R.: Androgen insensitivity syndrome. In: "The Intersex Child", Ed. Josso, N., Karger Publications, Basel, p. 71 (1981).

93. Manual, M., Katayama, K.P., Jones, H.W.Jr.: The age of occurrence of gonadal tumors in intersex patients with a Y chromosome. Amer. J. Obstet. Gynecol. 124, 293-300 (1976).

94. Amrhein, J.A., Meyer, W.J.III., Jones, H.W.Jr., Migeon, C.J.: Androgen insensitivity in man. Proc. natn. Acad. Sci. U.S.A. 73, 891-894 (1976).

95. Griffin, J.E.: Testicular feminization associated with a thermolabile androgen receptor in cultured human fibroblasts. J. clin. Invest. 64, 1624-1631 (1979).

96. Saenger, P., Levine, L.S., Pang, S., New, M.I.: Sexual ambiguity at birth. In: "Diagnosis in Andrology, Clinics in Andrology", Eds. Bailey, J., Hafez, E.S.E., Martinus Nijhoff Publication Company, The Hague, pp. 31-52 (1980).

97. Burstein, S., Grumbach, M.M., Kaplan, S.L.: Androgen responsiveness is important in the management of microphallus. Lancet II, 983-986 (1979).

98. Belgorosky, A., Rivarola, M.A.: Sex hormone-binding globulin response to human chorionic gonadotropin stimulation in children with cryptorchidism, anorchia, male pseudohermaphroditism, and micropenis. J. clin. Endocrinol. Metab. 54, 698-704 (1982).

99. Amrhein, J.A., Klingensmith, G.J., Walsh, P.C., McKusick, V.A., Migeon, C.J.: Partial androgen insensitivity. New Engl. J. Med. 297, 350-356 (1977).

100. Gilbert-Dreyfus, S., Sebaoun, C.A., Belaisch, J.: Etude d'un cas familial d'androgynoidism avec hypospadias grave, gynecomastie et hyperoestrogenie. Ann. Endocr. (Paris) 18, 93 (1975).

101. Lubs, H.A.Jr., Vilar, O., Bergenstal, D.M.: Familial male pseudohermaphroditism with labial testes and partial feminization. J. clin. Endocrinol. Metab. 19, 1110-1120 (1959).

102. Rosewater, S., Gwinup, G., Hamwi, G.J.: Familial gynecomastia. Ann. intern. Med. 63, 377-385 (1965).

103. Pang, S., Hotchkiss, J., Drash, A., New, M.I.: Microfilter paper method for 17α-hydroxyprogesterone radioimmunoassay. J. clin. Endocrinol. Metab. 45, 1003-1008 (1977).

104. Money, J., Hampson, J.G., Hampson, J.L.: Hermaphroditism: recommendations concerning assignment of sex, change of sex and psychologic management. Bull. Johns Hopk. Hosp. 45, 284-300 (1955).

105. Ehrhardt, A.A., Meyer-Bahlberg, H.F.L.: Effects of prenatal sex hormones on gender-related behavior. Science 211, 1312-1318 (1981).

106. Amrhein, J.A., Meyer, W.J.III., Danish, R.K., Migeon, C.J.: Studies of androgen production and binding in 13 male pseudohermaphrodites and 13 males with micropenis. J. clin. Endocrinol. Metab. 45, 732-738 (1977).

107. Savage, M.O., Chaussain, J.L., Evain, D., Roger, P., Job, J.C.: Endocrine studies in male pseudohermaphroditism in childhood and adolescence. J. clin. Endocrinol. Metab. 8, 219-231 (1978).

Subject Index

Bei Fragen zur Produktsicherheit wenden Sie sich bitte an:
If you have any questions regarding product safety,
please contact:

Walter de Gruyter GmbH
Genthiner Straße 13
10785 Berlin
productsafety@degruyterbrill.com